Macromolecular Systems – Materials Approach

Springer

Berlin
Heidelberg
New York
Barcelona
Budapest
Hong Kong
London
Milan
Paris
Santa Clara
Singapore
Tokyo

Michael Braden

R. L. Clarke · J. Nicholson · S. Parker

Polymeric Dental Materials

With 32 Figures, Some in Color and 34 Tables

 Springer

Professor Michael Braden
Dr. Richard L. Clarke
Dr. Sandra Parker
St. Bartholomews and the Royal London School of Medicine and Dentistry
Turner Street
London E1 2AD
United Kingdom

Dr. John Nicholson
Dental Biomaterials Department King's Dental Institute
King's College School of Medicine and Dentistry
Caldecot Road
London SE5 9RW
United Kingdom

Editors of the professional reference book series
Macromolecular Systems – Materials Approach are:
A. Abe, Tokyo Institute of Polytechnics, Iiyama
L. Monnerie, Ecole Superieure de Physique et de Chimie Industrielles, Paris
V. Shibaev, Moscow State University, Moscow
U. W. Suter, Eidgenössische Technische Hochschule Zürich, Zürich
D. Tirrell, University of Massachusetts at Amherst, Amherst
I. M. Ward, The University of Leeds, Leeds

CIP Data applied for
Die Deutsche Bibliothek – CIP-Einheitsaufnahme
Polymeric dental materials : with 34 tables / Michael Braden ...
– Berlin ; Heidelberg ; New York ; Barcelona ; Budapest ;
Hong Kong ; London ; Milan ; Paris ; Santa Clara ; Singapore ;
Tokyo : Springer, 1997
   (Macromolecular systems – materials approach)

   ISBN-13: 978-3-642-64450-4   e-ISBN-13: 978-3-642-60537-6
   DOI: 10.1007/978-3-642-60537-6

NE: Braden, Michael

Softcover reprint of the hardcover 1st edition 1997

Coverdesign: de'blik, Berlin
Typesetting: Fotosatz-Service Köhler OHG, Würzburg
SPIN: 10124628   2/3020 – 5 4 3 2 1 0 – Printed on acid-free paper

Preface

Oral disease is very rarely life-threatening, but affects nearly everyone over the age of five, certainly in the western world. In those countries where health care is undertaken by government, as in the UK, dentistry is very demanding on health service resources. Dental care costs in England and Wales approach £1 billion per annum; and most of this is on restorative dentistry, that is filling teeth and providing crowns, bridges, and full and partial dentures. The repair and adjustment of existing dentures alone costs around £17 million – approximately half the cost of orthopaedic implant revisions. Hence dental materials is a major part of the biomedical materials speciality and, indeed, has contributed its technology to other areas, such as orthopaedics, audiology and ophthalmics. In the UK, 1995 has seen two major reports on biomaterials, one emanating from the Materials Technology Foresight Programme and the other from the National Health Service, both from their own standpoints highlighting priority areas for future research. Both have identified dentistry as a priority area.

Before the second world war there were just two polymeric materials in general use, both natural products. These were Vulcanite ("Hard Rubber") for denture bases, which was highly cross-linked natural rubber, with soft vulcanised natural rubber occasionally used as a soft-lining material, and a form of Agar-Agar used as an "elastic" impression material. When Japan occupied much of the far east in 1942, these two materials were no longer available. The situation was restored by the development of alginate impression materials by the former Amalgamated Dental Company, and the adoption of poly(methyl methacrylate) (PMMA) as a denture base. Imperial Chemical Industries had already introduced PMMA as an injection mouldable material, but the "dough technique", patented by the Kulzer Company, was ultimately preferred due to the ease of processing. Poly(methyl methacrylate) has since been the main material of choice for full dentures, albeit with well known defects. This has, over the years, also posed the problem of providing soft lining materials for such dentures.

In the mid-1950s, fluid room temperature vulcanising (RTV) rubbers were developed, both silicone and polysulphide types. These were quickly adapted for dental use, transforming the dimensional accuracy of impressions; subsequently other modifications of this principle followed.

In the mid-1960s, D.C. Smith developed the first dental cement with chemical adhesion to dental hard tissues, based on zinc oxide and polyacrylic acid. The use of polyacrylic acid was extended by A.D. Wilson in developing a new generation of aesthetic filling materials, also using aluminosilicate glasses, used previously for many years in silicate cements. These were originally termed "glass ionomer cements" by Wilson, but are now generally referred to as "poly-alkenoate cements". Since their inception, these materials have undergone substantial development which shows no sign of abating. They have added a new dimension to restorative dentistry. At about the same time, the first of the polymeric composite filling materials were developed by R. Bowen in the USA. Such materials were the original alternatives to silicate cements as aesthetic filling materials, offering superior strength and resistance to oral fluids. This area itself has grown into a multi-million dollar industry, has also been the focus of intense developmental research, and again shows no sign of abating. In this case, the question has always been whether such materials could ultimately replace dental amalgam.

This uncertainty has in the last year or so been more sharply focussed by the ongoing controversy on the toxic effect on patients of mercury in amalgam. It is not intended to comment on this controversy here; suffice it to say that some European countries have already banned the use of amalgam for certain categories of patients. There seems, however, to be considerable social pressure for replacing amalgam, and composite materials seem to be the most likely candidate materials.

Hence the last 50 years has seen the proliferation of both the number and chemical sophistication of polymeric dental materials. This has involved the study of a wide range of material properties, some of which are common with materials as a whole, and others specific to dentistry. The continuing growth of regulatory requirements with respect to toxicity adds an extra dimension to the developement of dental materials. The following chapters in this book review the status of these materials.

A new area of polymeric dental materials is in the early stages of research, namely drug delivery for the treatment of periodontal disease, and other oral diseases. This is in response to the slowly changing pattern of dental disease, and specific needs such as oral disease in AIDS patients. It was decided that it is not timely to include this emerging area in this book, however promising it may be for the future.

London, Great Britain   *Michael Braden*
March 1997

Contents

Symbols and Abbreviations

C_p	Thermal Heat Capacity at Constant Pressure
C_s	Concentration of Photo-Sensitiser
D	Diffusion Coefficient
D_0	Pre-Exponential Term in Arrhenius Equation for Diffusion Coefficient
E	Young's Modulus
ΔE	Activation Energy
G	Shear Modulus (Real Part)
ΔG_{mix}	Free Energy of Mixing
ΔH_{mix}	Enthalpy of Mixing
I	Intensity of Radiation
K_T	Thermal Conductivity
K and α	Constants in the Mark-Houwink-Sakurada Equation
M_t	Mass Change At Time t
M_v	Viscosity Average Molecular Mass.
M_∞	Mass Change at Equilibrium
N_1	Number of Moles of Component $_1$
N_2	Number of Moles of Component $_2$
R	Universal Gas Constant
T	Temperature (degrees K)
X	Volume Fraction of Filler (%)
α	Linear Absorption Coefficient
α_s	Linear Absorption Coefficient of Photo-Sensitiser
ε_s	Molar Aborptivity of the Photosensitiser
η	Viscositys
η_{sp}	Specific Viscosity
$[\eta]$	Intrinsic Viscosity
ϱ	Density
Φ_1	Volume Fraction of Component $_1$
Φ_2	Volume Fraction of Component $_2$
Ψ	Wave Function
τ	Transmittance
γ	Absorption Coefficient
c	Concentration
d	Depth
k	Thermal Diffusivity
l	Semi-Thickness of Plane Sheet
Δs_{mix}	Entropy of Mixing
t	Time
$\tan \delta$	Mechanical Loss Tangent
x	Distance

CHAPTER 1

Polyelectrolyte Restorative Materials

J.W. Nicholson

1.1
Introduction

The history of polyelectrolyte restorative materials began in 1968 when Dr (now Professor) Dennis Smith, then of Manchester University, England, announced the invention of zinc polycarboxylate cement [1.1]. This material was made from poly(acrylic acid) by reaction with deactivated zinc oxide. The cement showed many exciting properties: it was bland towards oral tissues, set rapidly to show good resistance to oral fluids and, above all, it adhered to teeth. It thus ushered in the era of adhesive dentistry for, prior to its invention, dental restoratives had had to be retained mechanically by means of undercut cavities.

Following closely on the invention of zinc polycarboxylate came the glass polyalkenoate cement, known also by the trivial name of glass-ionomer. This material was invented by Alan Wilson and Brian Kent at the UK Laboratory of the Government Chemist, and also contained poly(acrylic acid). It thus shared with zinc polycarboxylate the properties of blandness and adhesion. By contrast with the zinc polycarboxylate, the base in this material was a special acid-decomposable glass. This glass was a multi-component material, based on calcium aluminofluorosilicate, but with other ingredients, such as aluminium phosphate and cryolite (Na_3AlF_6). Over the years an enormous range of glasses suitable for use in these cements has been made and their chemistry and phase relationships studied. Nonetheless, it is still fair to say that their composition and chemistry remains only partially understood.

The inclusion in the glass of fluoride-containing species conferred a particular advantage on glass-ionomers: the set cements release fluoride. This exchanges with OH^- ions in the hydroxyapatite of the surrounding tooth material to yield the more caries-resistant compound fluoroapatite (sometimes simply called apatite). Glass-ionomers are thus cariostatic restoratives, a feature that will be discussed in more detail later on.

The use of glasses in place of zinc oxide conferred other advantages over the zinc polycarboxylates. These include greater strength, due to the greater scope of varying the chemical nature of the cement matrix and its rate of formation, and greater translucency, leading to better aesthetics. These have allowed glass-ionomers to be used in novel clinical applications not accessible to other restorative materials, even zinc polycarboxylates. Such techniques include repair of erosion cavities at the gum line (so-called Class V cavities) and use as a dentine substitute in the biomimetic technique known as the *sandwich restora-*

tion. These advantages spring partly from the philosophical point once made by Alan Wilson: that the greater the scope for altering a material's composition the greater the scope for its improvement [1.2].

Glass-ionomers suffer, however, from a major disadvantage, namely that they are sensitive to the moisture content of their surroundings. If they dry out, the setting reaction ceases and they perform in a substandard way; on the other hand, if they are exposed to moisture for too long, the water-soluble cross-linking ions are washed out of the material and the result is also a substandard cement. Modern materials are less sensitive to this than were the earliest glass-ionomers on the market, though this remains a problem. Moreover, the much greater sensitivity of the early glass-ionomers to moisture gave this whole class of materials a reputation for being very sensitive to the clinical technique of the individual dental practitioner.

Partly in an attempt to overcome this early sensitivity to moisture, manufacturers have in the last few years developed resin-modified glass-ionomers. These materials are blends of conventional glass-ionomer components with organic monomers and photoinitiators, and generally possess the capability of being cured by irradiation with visible light at the blue end of the spectrum (470 nm). Early materials were launched as liners and bases, and were followed by restorative grades. More recently, luting cements based on resin-modified glass-ionomers have been developed.

Systematic studies have shown that these materials have some weaknesses not possessed by the original glass-ionomers: in addition to the inevitable polymerisation shrinkage associated with the setting reaction of the monomers, they have a tendency to swell in water. This is a result of the inclusion of hydroxyethyl methacrylate, HEMA, (see Fig. 1.1) in the formulations, both as co-solvent for the aqueous and organic constituents, and as (co)reactant for the photocure. The photocured cement contains either polymers or copolymers of HEMA, and these behave as hydrogels within the structure, leading to the sorption of water. This causes not only swelling but, in many cases, a marked reduction in strength.

However, it is still early days with these materials. They do offer considerable advantages to the clinician, both in terms of their so-called *command set* property, and because they are less moisture-sensitive than the self-cure glass-ionomers. Moreover, if one takes account of the point made earlier about greater scope for altering a material's composition implying greater scope for improvement, then these resin-modified glass-ionomers hold promise of enormous improvements in the future.

The current chapter covers all aspects of these materials, giving an account of both their structure and the effect that this has on their properties. All three of the above groups of materials are still the subject of research, both scientific and clinical, and the present debates on these materials are reflected in the sections

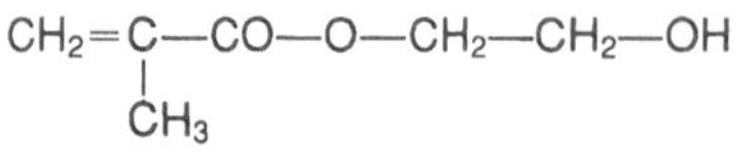

$CH_2\text{=}C\text{—}CO\text{—}O\text{—}CH_2\text{—}CH_2\text{—}OH$
$\quad\ \ |$
$\quad\ CH_3$

Fig. 1.1. Structure of 2-hydroxyethyl methacrylate

that follow. First, though, two topics fundamental to the understanding of these materials are considered, namely the chemistry of polyelectrolytes themselves, and then the rôle of water in the cements.

1.2
Polyelectrolytes

Polyelectrolytes are linear polymers having a multiplicity of ionisable functional groups. In solution they dissociate into polyions and small ions of opposite charge, known as counterions. The polyelectrolytes used in dental restorative materials are all anionic, i.e. the polymer carries negative charges. The counterions are thus positively charged.

The main functional groups found in dental polyelectrolytes are carboxylic acids, though recently phosphonic acids have been described for this purpose [1.3]. Sulphonic acid polymers, though capable of forming cements, do not yield materials of sufficient hydrolytic stability for use in dentistry.

Polyelectrolytes may be homopolymers or copolymers. The homopolymer of acrylic acid is widely used in both zinc polycarboxylate and glass-ionomer cements, but copolymers, such as acrylic/maleic and acrylic/itaconic acids are also used. To date, of the phosphonic acids, only homopolymers of vinyl phosphonic acid have been used, though copolymers of this monomer are known.

A feature that arises from the high charge density is that polyelectrolytes are soluble in water. Aqueous solutions of polyelectrolytes behave quite differently from either solutions of uncharged polymer in organic solvents or from low molar mass electrolytes. This is because of the effects arising from the combination of properties that result from interactions of electrical charges [1.4]. Unfortunately, these effects have generally been studied under conditions far removed from those employed in dental cements, typically very dilute solutions and relatively high added salt concentrations. Dental cements, by contrast, typically employ concentrated solutions, of the order of 40–50 mass%, and have no other electrolytes present [1.5]. However, some consideration of the results from dilute solutions is appropriate at this point, since some of the conclusions are relevant when extrapolated to regions of high concentration.

Anionic polyelectrolytes tend to adopt helical conformations in aqueous solution [1.6]. This conformation, which prevails at the naturally low pH (i.e. 1.5–2 for poly(acrylic acid) depending on concentration and molar mass) is stabilised because it results in minimum electrostatic repulsion. However, progressive neutralisation of such a polyelectrolyte results in gradual increase in charge density, which in turn leads to progressive coil expansion. In dilute solution, this coil expansion continues until the molecule is no longer helical but rod-like. The helix-to-rod transition appears fairly well defined and can be determined inter alia by changes in viscosity.

In reducing the polyelectrolyte concentration, similar effects can be seen, in particular that dissociation of the acid functional groups becomes more extensive with increasing dilution [1.5]. In order to determine the molecular weight from viscosity, a graph of η_{sp}/c vs c is plotted, and the line extrapolated to c = 0 to determine $[\eta]$, the reduced viscosity. The reduced viscosity is then used in the Mark-Houwink-Sakurada equation, together with previously determined values of K and α, to evaluate M_v, the viscosity-average molecular weight:

$$[\eta] = KM_v^\alpha$$

However, for polyelectrolytes, this procedure cannot be adopted: the increased dissociation with increasing dilution results in an upswing in the slope of the plot of η_{sp}/c vs c, hence extrapolation to zero is impossible. This is sometimes referred to as the polyelectrolyte effect.

In order to overcome the polyelectrolyte effect, a simple 1:1 electrolyte, such as sodium chloride, is added to the solution. This has the effect of screening the electrostatic interactions in the polymer and of eliminating the concentration dependence of the degree of dissociation. For this reason, viscosity-average molecular weight determinations are carried out in aqueous electrolyte solutions and K and α values are quoted for solutions of defined concentration of NaCl or NaBr. The level of added salt is important, and must exceed the equivalent concentration of functional groups on the polyelectrolyte if the polymer is to behave as though it were uncharged. This level is sometimes referred to as the *abundant salt* region [1.4] and it results in counterions clustering in the zones immediately adjacent to the polyelectrolyte backbone. This is called *condensation* and the ions are described as *condensed* [1.6].

There are problems with the level of salt added: very little is known about the situation in which the concentration of polyelectrolyte functional groups exceeds that of the added salt [1.4], though there is some not very extensive data on the solution properties of polyelectrolytes alone, with no added salt. Theory for this latter situation is very conjectural, and the interpretation of the experimental data more complex than for the abundant salt regime. Conversely, the situation of the abundant salt regime is itself complex at higher concentrations, since the addition of large concentrations of NaCl or NaBr causes substantial amounts of water to become involved in co-ordination around the ions, and this affects solvent quality. At high enough concentrations, this may lead to the Θ (theta) condition [1.4].

Polyelectrolyte molecules adopt two conformations: the random coil and the ordered helix. In the latter, the structure contains regular repeating units along the molecule; conversely, in the former, it contains no regular features, and hence the term *random* coil.

Random coil conformations can range from the spherical, contracted state to the fully extended rod-form, depending on charge and concentration [1.7]. Though there is little data on highly concentrated solutions of polyelectrolytes as used in dental cements, what evidence there is suggests that they tend to adopt relatively contracted and spherical conformations. By contrast, under more dilute conditions, they undergo considerable expansion as charge is progressively increased. For poly(acrylic acid) with a degree of polymerisation of 1000, neutralisation causes an expansion from a spherical conformation of diameter 20 nm to a rod of length 250 nm [1.8].

The extent to which this kind of conformational change can occur in concentrated polyacid solutions used in dentistry might be thought to be limited.

Table 1.1. Effect of progressive neutralisation with NaOH on the properties of poly(acrylic acid) (from Wasson [1.9])

Neutralization (%)	pH	Density (g/cm³)	Viscosity (cP)
0	1.38	1.15	100
10	3.16	1.17	409
25	3.83	1.19	907
40	4.33	1.24	2210
55	4.71	1.27	4334
70	5.30	1.33	5330
85	5.84	1.35	13097
100	7.83	1.36	51552

However, the effect of neutralisation is, in fact, considerable, as the data in Table 1.1 show. The development of fully formed negative charges in the polyelectrolyte functional groups clearly causes conformational changes that lead to increased viscosity of several orders of magnitude. This feature is consistent with coil expansion and increased chain entanglement, albeit that the coil expansion is probably less than would occur in more dilute solutions, and almost certainly does not result in the development of a completely rod-like morphology.

Neutralisation of polyelectrolyte cements has the added complication that it occurs accompanied by ion binding. This means that counterions become strongly associated with the polyanion chain, an effect which has the opposite result from neutralisation in than the charge density in the region of an individual polymer molecule is reduced.

Ion binding as a phenomenon has been demonstrated to occur by a number of techniques, including titration [1.10, 1.11], viscosity and electrical conductance measurements [1.12, 1.13], dilatometry [1.14, 1.15], and NMR spectroscopy [1.16, 1.17]. It has been shown to be affected by the size and charge of the counterion, the charge and conformation of the polyanion, and the states of hydration of the participating species.

Oosawa [1.8] has given a detailed account of the distribution of counterions around a polyion. To begin with, he has identified three possible states for the counterions: free (i.e. not associated at all with any of the polyions), bound but relatively mobile (termed *atmospheric*), and tightly bound to individual sites on the polymer (termed *site-bound*). The atmospheric ions occupy the potential valley around each polyion, a cylindrical region, defined by the shape and dimensions of the polyion, but extending beyond it by a defined amount in all three axes of space. Bound ions exist in intimate association with specific functional groups, effectively filling potential holes created by the existence of charges on the individual functional groups, and resulting in the formation of an ion-pair.

Site-binding of ions may be assisted by partial covalency of the participating ions. For example, divalent cations, such as Mg^{2+} and Zn^{2+} undergo some polarisation under the conditions prevailing in aqueous solutions of organic acids, and this results in the partial covalency, one effect of which is the development of some directional character in the interaction of counterion and polyion. Studies of complexation constants of divalent metals with 0.06 N aqueous poly(acrylic acid) showed the order of stability to be [1.15]:

Mg < Ca < Co < Zn < Mn < Cu

As Wilson and Nicholson [1.5] have observed, some of these divalent metal ions form part of the Irving-Williams series which, irrespective of the nature of the co-ordinating ligand, runs [1.18]:

Mn < Fe < Co < Ni < Cu > Zn

Hence, site binding in metal polyacrylate complexes largely follows the pattern of the Irving-Williams series, as do the stabilities and strengths of polyelectrolyte cements derived from poly(acrylic acid) and the appropriate divalent metal oxide.

The extent of ion binding depends not only on the nature of the cation, but also of the polyion. Its degree of dissociation, acid strength, conformation, distribution of ionisable groups and hydration state all affect ion binding [1.8, 1.19]. For example, alkali metal ions have been shown to bind more strongly to poly(acrylic acid) than to poly(methacrylic acid), an effect which correlates with pH of the parent polyacid [1.19]. The strength of the ion binding is enhanced by the possibility of the formation of chelated structures of carboxylate groups with the ion. For example, Mg^{2+} has been shown to bind more strongly to poly(vinyl methyl ether-maleic acid) than to either poly(acrylic acid) or poly(ethylene-maleic acid).

Water molecules become strongly oriented in strong electrostatic fields, such as those existing in the regions of atmospheric or site-bound ions. This oriented water is different from ordinary water: it has a higher refractive index and greater density [1.15, 1.20, 1.21]. The structure of this water is affected by ion binding, and can be determined by measuring changes in refractive index or density. The rôle of water in these materials is considered in more detail in the next section.

Ikegami [1.20, 1.21], for example, used changes in refractive index to determine the state of water in polyelectrolyte-metal complexes. He showed that, at low degrees of neutralisation, the average distance between the ionised groups is great, and that rearrangement of water molecules is due solely to the charge on the individual functional group. Individual hydration spheres of oriented water, so-called *intrinsic water*, are formed at each charged site: in the case of poly(acrylic acid) at 30% neutralisation, these spheres are of radius 0.31 nm [1.21]. As neutralisation proceeds, the distance between the ionised groups decreases and there begins to emerge a co-operative effect. Above 30% neutralisation the discrete water spheres begin to coalesce and eventually form a cylindrical sheath around the polyanion. With further increases in degree of neutralisation, a second cylindrical sheath of water develops, in which water molecules are oriented by the effect of two or more carboxyl groups. At full neutralisation, the primary sheath has a diameter of 0.5–0.7 nm [1.21], while the second, outer cylindrical hydration region has a diameter of 0.9–1.3 nm.

In addition, the formation of a stable hydrogen-bonded ring structure as in poly(itaconic acid) and in poly(maleic acid) has also been shown to influence hydration states [1.22, 1.23].

1.3
The Rôle of Water

The setting reaction of polyelectrolyte restorative materials takes place in water, but not in such a way that phase separation occurs. Instead, all of the water originally included in the initial cement paste becomes included in the hardened material. For example, glass-ionomer cements typically include at least 15 mass% of water within the initial formulation, all of which becomes incorporated into the set cement [1.24].

In addition to its inclusion as such in the initial cement pastes, water is one of the products of the acid-base setting reaction. Hence the neutralisation reaction generates yet more water, and this, too, is retained within the set cement. Water is thus both solvent for the setting process and a component of the hardened cement.

Water may have a number of rôles in set polyelectrolyte cements. It may solvate the cement-forming ions, such as Ca^{2+} or Zn^{2+}, depending on the cement. It is present as a sheath solvating the actual polyelectrolyte, such as poly(acrylic acid). Water is known to be retained by metal poly(acrylate) salts at equilibrium, such entrained water acting as plasticiser and lowering the glass transition temperature of the polysalt [1.25].

The ions that form cements with polyelectrolytes are such species as Al^{3+}, Mg^{2+}, Ca^{2+} and Zn^{2+}, all of which are capable of developing the co-ordination number 6, and for each of which well-characterised hexaquo ions are known [1.26]. Most of these ions fall into the category of *hard* in Pearson's Hard and Soft acids and Bases (HSAB) scheme [1.27]. The principle of this scheme is that bases may be divided into two categories, namely those that are readily polarised (*soft*) and those that cannot be polarised (*hard*). Lewis acids (i.e. chemical species that act as electron pair acceptors) can also be divided into hard and soft groups, depending on polarisability. The useful generalisation then emerges that hard acids prefer to associate with hard bases, and soft acids with soft bases [1.27].

Of the ions mentioned, Al^{3+}, Mg^{2+} and Ca^{2+} are definitely hard, while Zn^{2+}, by contrast, falls into the category designated by Pearson as *borderline*. This means that most of these ions form their strongest complexes with hard bases, i.e. those involving non-polarisable elements from the first row of the periodic table, such as oxygen and nitrogen. Water is thus a hard base, and the complexes that it forms with these ions involve co-ordination by the oxygen atoms. As predicted by the HSAB concept, the aquo complexes of the cement-forming ions are highly stable and hence do not readily lose the co-ordinated water.

Water occurs in both zinc polycarboxylate and glass-ionomer cements in at least two different states, classified as evaporable (unbound) and non-evaporable (bound). This is based on whether or not the water can be removed by vacuum desiccation over silica gel or other similar dehydrating treatment [1.5]. In the glass-poly(acrylic acid) system the evaporable water is up to 5 wt% of the total cement, while the bound water is $18-28$ wt% [1.28]. This amounts to bound water at a level corresponding to five or six water molecules for each acid group and each associated metal cation (i.e. at least ten water molecules per metal carboxylate site). This contrasts with the zinc polycarboxylate cement, where bound water is present at a level corresponding on average to slightly under two molecules per zinc ion [1.29], and most of the water in the cement is unbound.

In polyacrylate molecules, the individual carboxylic acid groups have been assumed to be surrounded by a primary local sphere of oriented water molecules [1.20]. In addition, a secondary sheath of water molecules surrounds the whole molecule. This secondary sheath is maintained in place as a result of the co-operative action of the polar functional groups along the polymer backbone. Monovalent ions such as Li^+ or Na^+ are able to penetrate only this secondary sheath. Hence they form a solvent-separated ion pair. Divalent ions, by contrast, cause a much greater disruption to both the secondary hydration sheath and the primary water layers. Gelation of poly(acrylic acid) by ions such as Mg^{2+} or Ba^{2+} has been assumed to arise due to complete displacement of water molecules from around the carboxylic acid sites, i.e. effectively causing dehydration of these sites [1.30]. Such gelation phenomena are clearly likely with the divalent ions available in polyelectrolyte dental cements.

In addition to the possibility of close approach by divalent ions to carboxylic acid sites on the polymer, their preferred co-ordination behaviour predisposes them to complex with oxygen atoms of the carboxylate groups. This means that divalent ions are able to become site-bound to individual ionised acid groups within the cement. Such behaviour contrasts with that of monovalent ions, which are known not to become site bound, but to remain labile and capable of hopping between carboxylate groups [1.31]. Divalent ions thus form salt-like bridges which may act as crosslinks between the polymer molecules [1.32].

1.4
Zinc Polycarboxylate Cements

1.4.1
History and Clinical Uses

Zinc polycarboxylate cement consists of a modified zinc oxide powder, heat treated to make it very slightly non-stoichiometric, and poly(acrylic acid), originally in aqueous solution [1.1]. More recently, versions have become available in which the zinc oxide powder is combined with a dried powder of poly-acid, and the whole mixture activated by the addition of water. Not all of the zinc oxide powder reacts, so that, when set, the cement consists of a zinc polyacrylate matrix with unreacted particles embedded in it as reinforcing filler.

Zinc polycarboxylate cement is now widely used for such clinical procedures as lining cavities prior to placement of the main filling material, for attaching crowns to posts, and for the adhesion of orthodontic brackets as part of the treatment for misaligned teeth. Typical properties of zinc polycarboxylate cements are given in Table 1.2.

Table 1.2. Properties of zinc polycarboxylate dental cement

Property	Typical components/values
Liquid	40–50% polyacid
Polymer	Usually poly(acrylic acid)
Powder: liquid ratio	2.5–3: 1
Setting time/min	2.5–4
Compressive strength/MPa	80–100

1.4.2
Structural Studies

The zinc oxide used for these cements is modified for the purpose by mixing pure zinc oxide with small amounts of magnesium oxide and fusing the mixture at between 1100 and 1200 °C. This process reduces the reactivity of the zinc oxide towards the acid so that, in clinical use, the cement paste sets slowly enough to be mixed and placed. The heat treatment causes the zinc oxide to become slightly yellow in colour. This coloration is due to evaporation of oxygen to yield a non-stoichiometric substance corresponding to $Zn_{(1+x)}O$, where x is less than or equal to 70 ppm [1.33].

Studies on the structure of the cement are still continuing. For example, Nicholson et al. [1.34] have shown, using Fourier transform infrared spectroscopy, that, within the cement, zinc ions are partly chelated to the carboxylate groups of the polymer. The structure of these units is complex, with a range of types with subtly different chelating character being present as illustrated in Fig. 1.2. Covalent structures of this type are also known to occur in a variety of monomeric zinc carboxylates [1.35].

The finding that carboxylate units had some covalent character contrasts with previous claims that this material is purely ionic, based on the use of the older dispersive infrared spectroscopy. In fact, this technique lacks the resolving power of FTIR and therefore cannot readily distinguish between ionic and covalent structures.

In another study, Hill and Labok [1.36] showed that these cements behave, in many ways, like thermoplastic composites with very weak crosslinks between the chains. This study, which involved determining the relationship between the molar mass of the polymer and the fracture toughness of the cement, supported conclusions obtained some years previously concerning the relatively plastic nature of this cement [1.37].

The rôle of water has been considered by, among others, Nicholson et al. [1.29], who studied the effect of desiccation and of soaking in water and 0.9% saline solution on the compressive strength of specimens of zinc polycarboxylate cement. These cements were shown to be relatively poorly hydrated, particularly by comparison with the closely related glass-ionomer cements. Water seems to occupy co-ordination sites around the zinc ions, but does not have any other structural rôle. Little or no differences were seen in the hydration, setting or strength properties of the materials, regardless of

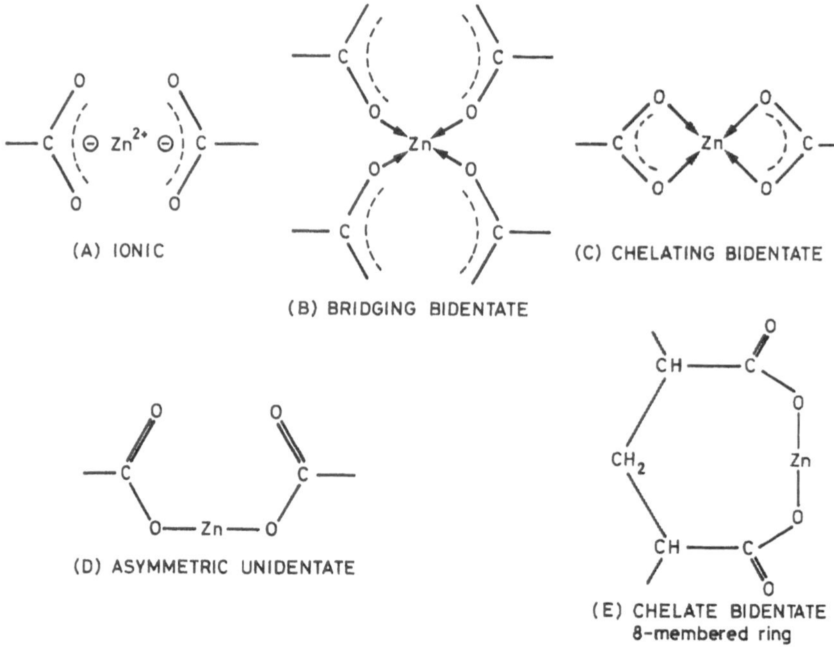

Fig. 1.2 A–E. Possible structures of zinc carboxylate units

whether they were made from aqueous acid with zinc oxide or from a water-activated mixture of polyacid plus ZnO.

1.4.3
Studies on Adhesion

Zinc polycarboxylates are widely used as adhesives in dentistry, but experimental studies usually report considerable scatter in the values for the adhesive strength of bonded joints. The probable source of this scatter is lack of control of the adhesive layer thickness, a parameter studied experimentally by Akinmade and Hill [1.38]. In their paper, they reported the effect of varying the adhesive layer thickness for both a zinc polycarboxylate and a glass-ionomer cement, using a linear elastic fracture mechanics approach to analyse their data. They found that there was an optimum adhesive thickness for the zinc polycarboxylate corresponding to a plastic zone of about the same size, and that the plastic zone size varied with molecular weight of the poly(acrylic acid) from 40 μm at M_w of 1.15×10^4 to 290.9 μm at M_w of 3.83×10^5. By contrast, the glass-ionomer did not exhibit an optimum thickness, but since its calculated plastic zone size varied only between 5.7 and 10.5 μm over the same range of polymer molecular weight, this was not surprising. The work of Akinmade and Hill not only established the existence of the optimum adhesive layer thickness for zinc polycarboxylate cement, but also established the large differences between the plastic zone size in it and the apparently closely related glass-ionomer cement.

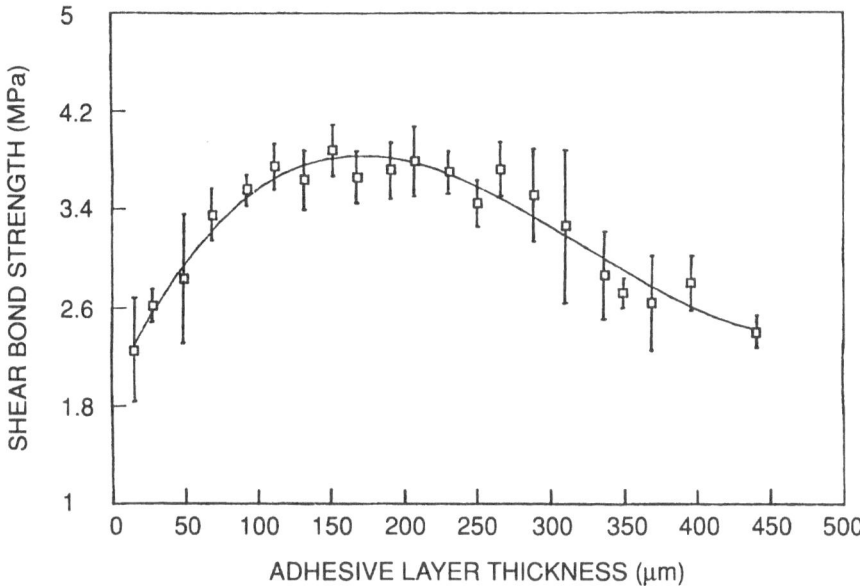

Fig. 1.3. Plot of shear bond strength against adhesive layer thickness for zinc poycarboxylate dental cement, showing maximum at 205 μm (after Akinmade and Nicholson (1995) *Biomaterials,* 16:149–153)

In order to control the thickness of the layers, Akinmade and Hill used glass ballotini as spacers when the sample was cured between two glass plates. This technique can be criticised on the grounds that it introduced a foreign material into the cement, one that was likely to inhibit the setting reaction and in other ways alter the properties of the set cement. To overcome this objection, further work was carried out [1.39] using an alternative method of varying the adhesive layer thickness; in addition a commercial zinc polycarboxylate dental cement was used. A specially designed jig was built in which the thickness of the adhesive could be varied by altering the spacing of stainless steel test pieces in a controlled way prior to bonding. The use of this apparatus removed the need to include glass spacers within the cement itself. The test pieces fabricated in this way were in the form of lap joints that were then tested to failure in shear. Using this apparatus the existence of an optimum thickness was confirmed for the commercial brand of zinc polycarboxylate dental cement, PolyF Plus (DeTrey Dentsply). This optimum thickness was 205 μm, a value that is in good agreement with the earlier findings for the experimental zinc polycarboxylate cements. These results, which are shown in Fig. 1.3, also indicate that the commercial material has a plastic zone size comparable to those in the experimental cements derived from polyacids of similar molar mass, i.e. in the region of 200 μm. In addition, the latter work showed that thicker layers of zinc polycarboxylate are more likely to fail cohesively than thinner ones.

1.5
Glass-Ionomer Cements

1.5.1
History and Clinical Uses

The development of the glass-ionomer cement arose from previous fundamental studies on dental silicate cements [1.40] and studies of cements made from organic acids with chelating character [1.41]. It was assisted by work on the zinc polycarboxylate cement [1.1] in which Smith showed that dental cements exhibiting the property of adhesion could be prepared from poly(acrylic acid). Glass-ionomers have therefore been described as hybrids of dental silicates and zinc polycarboxylates. However, as Wilson and McLean have pointed out [1.24], neat as this view seems, it is not actually correct. To make a practical cement requires that the reaction between the aqueous polymer solution and the glass take place in a controlled fashion, and this formidable problem was solved only by developing new glasses of altered basicity and, slightly later, by the discovery of the effect of the additive (+)-tartaric acid on the system. Moreover, recent research has highlighted just how different glass-ionomers are from zinc polycarboxylates.

The initial problem was to enhance the reactivity of the glass towards the polymer. Early cements set sluggishly, had short working times (as opposed to setting times), and remained sensitive to moisture loss and water ingress for extended periods of time after setting. The glass that first approximated to an acceptably reactive component, so-called G200, was high in fluoride, and itself highly opaque. It thus gave cements whose translucency was very low and hence which were unacceptable aesthetically.

The early difficulties began to be solved following the discovery of the effect of (+)-tartaric acid [1.42]. This material, added in small amounts, improved the manipulation of the cement paste by modifying the setting rate. As the data in Table 1.3 shows, it extended the working time and sharpened the set, and thus yielded a material with significantly enhanced properties [1.43]. So, too, did citric acid but, unlike (+)-tartaric acid, it caused an increase in the solubility of the resulting cement and hence was not acceptable as an additive in these cements. Despite the length of time since the discovery of its effect, the precise mode of action of (+)-tartaric acid remains unclear, even today.

Clinically, the use of glass-ionomer cements was developed to exploit the adhesion of the cement to natural tooth material. An early suggestion was the

Table 1.3. The effect of chelating additives on the properties of glass-ionomer cements (after Crisp et al. [1.42])

Additive	Level	Working time/min	Setting time/min	Compressive strength/MPa
None	–	2.5	6	165
(+)-tartaric acid	5%	2.0	4	181
Citric acid	5%	2.0	3.75	192

use in fissure sealing and filling [1.44], an application which also exploited the fluoride-releasing capability of the material. Other suggestions to emerge early in the development of these materials included their use in paediatric dentistry, as restoratives for Class V erosion lesions, as a liner/base beneath composite resins and in minimal cavity restorations [1.45–1.47]. This latter idea was extended to develop the so-called *tunnel* preparation [1.48], in which caries is removed from within the tooth while preserving the maximum amount of surface enamel. This is followed by injecting glass-ionomer cement, which sets and acts as the core which binds the enamel shell together.

Glass-ionomers retain the characteristic of early sensitivity to moisture, though great improvements have been made to this since the earliest materials became available in the mid-1970s. Clinically, in the use of these materials, this feature still has to be taken into account.

Modern glass-ionomers undergo a sharp set approximately 4 min from the start of mixing. This is the time at which the matrix can be removed and the placement examined for defects. The material is still, however, slightly susceptible to water uptake and/or loss. To overcome this problem, a newly placed cement should be painted with a waterproof sealant as the matrix band is removed [1.49].

Manufacturers typically supply a special varnish sealant, but this does not necessarily provide a sufficiently impermeable finish, at least as just a single coat. An alternative procedure adopted by some clinicians is to employ a low viscosity light-cured unfilled resin system. As the matrix is removed, a generous amount of such resin is allowed to flow over the cement restoration. When the contour is completed, further resin can be added if necessary, and the whole protective coating cured by irradiation with light. The resinous layer can be removed after 24 h, at which time the cement can be finished by polishing under air/water spray. This technique gives optimum translucency and physical properties [1.49].

1.5.2
Clinical Manipulation

This subject has been covered in detail by Wilson and McLean [1.24] and by Mount [1.49], in both cases with extensive colour illustrations showing various techniques and step-by-step clinical procedures. Mount also discusses clinical handling in great detail. He points out, for example, that because glass-ionomers are water-based, they are always susceptible to further loss or uptake of water. Therefore, he recommends that both powder and liquid bottles be stored firmly closed. Where the liquid component is an aqueous polymer solution, such care will prevent water loss, with its attendant thickening and increased difficulty in mixing. Increased concentration leading to thickening may also alter the amount of solution dispensed in a single drop, leading to critical changes in the ratio of ingredients monitored at the chairside in terms of number of drops of liquid to number of scoops of powder.

Other brands of cement are presented in water-activated form, with powder consisting of glass mixed with the appropriate amount of dried poly(alkenoic acid). The liquid component for such cements is either pure water or an aqueous solution of (+)-tartaric acid. In this case, the powder is hygroscopic, and will take up water to clump and undergo a small amount of localised setting unless exposure to the air is kept to a minimum. Hence, for this type of cement, too, it is important to keep the bottles firmly closed between use [1.49].

Mixing is critical for the success of glass-ionomer cements. For those versions designed for hand mixing, the manufacturers generally supply a pad of waxed paper on which to perform the mixing operation. The difficulty with this approach is that water can soak into the paper, thereby altering the powder: liquid ratio, unless mixing is performed immediately following measuring out of the appropriate number of drops and scoops. The use of a glass slab as an alternative to the paper is recommended for two reasons: it will not affect the water balance and it can be chilled slightly in the refrigerator in order to extend the working time slightly. Whatever surface is used on which to carry out mixing, Mount [1.49] recommends a two-stage process in which the powder is divided into two approximately equal amounts, the first of which is folded into the liquid in 10–15 s, and the second of which is mixed into the resulting thin paste within the next 15 s.

The mixing of capsulated glass-ionomers may present practical problems for the clinician at the chairside. Although some manufacturers supply a purpose-built mixing machine, others rely on a capsule design that can be fitted into a high-energy amalgamator, and mixed by this machine. However, such amalgamators are known to vary widely in the amount of energy imparted to the mixture, depending on the details of their design, and even such local factors as power supply and power surges. These differences lead to differences in the efficiency of the mixing process. Mount suggests that clinicians test the efficiency of their amalgamators by determining the time to "loss of gloss" of the particular glass-ionomer cement [1.49].

This relies on the fact that as a glass-ionomer sets it undergoes a number of changes, including an alteration in appearance from wet and glossy to dull-looking and matt. At this point the setting cement has a detectable structure, in that it does not slump following probing with a spatula or pointed instrument. Determining this time to loss of gloss gives the clinician a useful means of measuring the approximate working time in the absence of the kind of laboratory apparatus used by researchers. Mount suggests that the working time may be estimated by determining the time to loss of gloss and subtracting 15 s. With this estimation technique, it is possible for the clinician to determine the effectiveness of mixing for the amalgamator or other mixing apparatus available at the chairside.

1.5.3
Fluoride Release

Glass-ionomer cements are known to release fluoride in a sustained manner, as has been demonstrated in numerous laboratory tests [1.50–1.52]. Under static conditions, release can be detected for at least 18 months [1.53], whereas under dynamic conditions (i.e. exposure to a continual flow of water) release has been detected for at least 5 years [1.54]. In the former case, at least, release appears to be a combination of initial washout, diffusion and erosion.

The effect of this fluoride is clinically beneficial, as has been demonstrated by Tyas in a five-year clinical study [1.55] and by Forsten [1.56]. Fluoride appears to be effective because it is taken up by the adjacent tooth material by an ion-exchange mechanism in which F^- is exchanged for OH^- in the hydroxyapatite of the enamel of the tooth. The effect is to transform the mineral phase to fluor-apatite, a material of greater stability towards acid attack than the native hydroxyapatite.

The transformation of the mineral phase of the adjacent tooth to fluorapatite has other beneficial effects. For example, the surface energy of the tooth is altered so that caries-promoting plaque is less able to adhere [1.58]. Fluoride also increases remineralisation of teeth [1.59], and reduces the fermentation of carbohydrates and the growth of plaque bacteria [1.60, 1.61]. Overall, then, the fact that glass-ionomers release fluoride is widely recognised as clinically advantageous.

1.5.4
Test Methods

The test methods for studying a range of acid-base cements, including glass-ionomers, have been comprehensively described by Wilson and Nicholson [1.5]. The methods range from spectroscopic, such as infrared and nuclear magnetic resonance (NMR), to conventional mechanical testing, such as determination of compressive and flexural strengths, and include physical tests such as for opacity. They also include a wide range of more specialised techniques, such as for determining working and setting times, and for determining consistency of the setting cement pastes.

Infrared spectroscopy has been useful for studying glass-ionomer cements because of the ease with which it can be applied, usually in the attenuated total reflectance (ATR) mode, to these materials. The region of interest is broadly the range $1550-1620$ cm^{-1}, which is where the asymmetric stretch of the carboxylate occurs [1.34, 1.62]. The exact position depends on both the nature of the chemical bonding involved, varying depending on whether such bonding is purely ionic or partially covalent, and being altered by different types of chelation by the carboxylate group [1.62]. This technique has been applied to setting studies of glass-ionomers [1.5], zinc polycarboxylates [1.63] and other metal oxide-poly(acrylic acid) cements [1.64].

More recently, Fourier Transform Infrared (FTIR) spectroscopy has become available, leading to greater sensitivity, and very much more rapid scan times, which in turn enhances the capability of the technique to give meaningful results on the setting reactions occurring in polyelectrolyte cements. FTIR has been applied to the setting chemistry of both zinc polycarboxylates [1.34] and glass-ionomer cements [1.5]. The technique showed for the first time that zinc polycarboxylate becomes partially covalent with time after setting. For glass-ionomers, the rôle of (+)-tartaric acid was shown to be to suppress the early formation of calcium polyacrylate and to enhance the formation of aluminium polyacrylate [1.5].

NMR spectroscopy has proved less useful in the study of these cements, largely because it is unable to record spectra fast enough to give meaningful results on setting processes. In an attempt to overcome this, Prosser et al. [1.65] used a much more dilute, hence slow setting, system than would be employed clinically. Applying ^{13}C NMR spectroscopy to the resulting cement showed for the first time that (+)-tartaric acid reacts preferentially with calcium ions released from the glass, and hence suppresses calcium polyacrylate formation.

Strength has been determined using a wide variety of configurations. Compressive strength is the most widely used of these techniques and, despite difficulties due to the complexity of failure in compression [1.5], it remains the method of choice. It is employed in the ISO standard test [1.66] and is sensitive to changes in the structure of the test materials. For example, it revealed the post-hardening maturation processes in glass-ionomers, even though these were not apparent from flexural tests, since early flexural strength showed little or no change on ageing [1.5]. Specimen geometry and loading rate alter results obtained, and ISO standard conditions employ 6 mm (high) × 4 mm (diameter) specimens that are loaded at 1 mm/min [1.66]. These dimensions are less than those used previously, prior to 1991, which were 12 mm (high) × 6 mm (diameter), but give results which are the same within an acceptable statistical variation, though standard deviations tend to be slightly greater for the smaller specimens [1.67]. Also, preparing specimens by building them up in layers, as is necessary for the resin-modified versions of the glass .onomer, has been shown to have a significant weakening effect [1.67].

Setting and rheological properties are important in the successful clinical use of polyelectrolyte cements. However, they are difficult to measure in the laboratory. For this reason, recourse is usually made to pragmatic tests which measure a related property in a readily determined way. Setting, for example, has been measured by determining the resistance to indentation by a weighted needle. This is based on a test originally devised in 1864 by Gillmore for studying the setting of hydraulic cements [1.68]. Though still in use, the Gillmore needle clearly does not measure anything except resistance to indentation but this has nevertheless been taken to be the criterion of setting in many published studies, and remains the method recommended in the current ISO standard for glass polyalkenoate dental cements [1.66].

A further test incorporated into the current ISO standard is the impinging jet acid erosion test. In this test, constant jets of liquid impinge onto the surface of the test specimens, as illustrated in Fig. 1.4.

The constant flow is maintained by a constant head device which feeds eight separate jets of 1 mm internal diameter with a recirculating pump and a reservoir of approximately 10 l capacity. The flow of liquid is 120 ± 4 ml/min, and is controlled by adjusting the height of the reservoir head. Specimens are held 10 mm below the jet of liquid. The liquid consists of buffered lactic acid at 20 mmol/l concentration and pH 2.7 at 23 °C. Specimens are subjected to the test for 24 h after mixing, and the test is run until the surfaces of the specimens have eroded by between 0.02 and 1.5 mm, the length of time to achieve this varying with the type of cement. Glass-ionomers typically require 24 h to erode by such an amount [1.66].

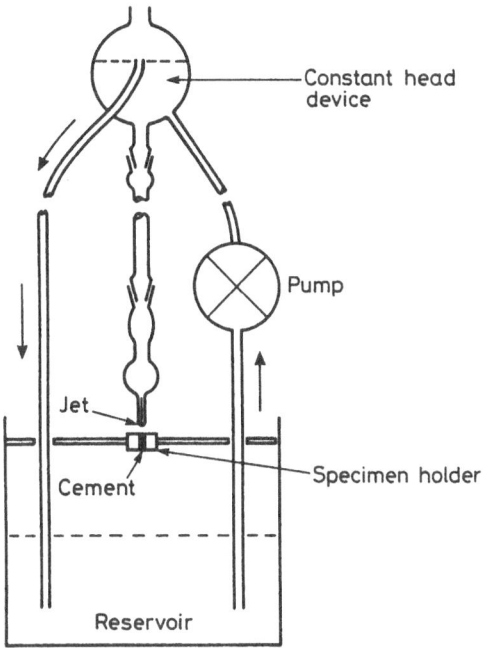

Fig. 1.4. Diagram of the impinging jet acid erosion test apparatus

Constant head device

Pump

Jet

Cement

Specimen holder

Reservoir

Table 1.4. Requirements of glass polyakenoates [1.66]

Cement type	Setting time/min	Compressive strength (MPa)	Acid erosion mm/h (maximum)
Liner/base	2–6	70	0.05
Restorative	2–6	130	0.05

The properties of glass-ionomers required by the current Standard are shown in Table 1.4.

1.5.5
Glasses

The powders used in glass-ionomer cements resemble the glasses used in the now obsolete dental silicate cements in that they are generally calcium aluminosilicates that are fused at high temperatures, typically in the range 1200 to 1550 °C [1.69]. Following fusion, the molten glass is shock cooled by pouring it onto a cool metal plate and then into water. This generates a coarse frit which is further ground, usually by dry milling in a rotary ball mill, until it is capable of passing through a 45 μm sieve (for a filling cement) or a 15 μm sieve (for a fine-grained luting cement). The ground glass powders may be annealed by heating at 400–600 °C, which reduces their reactivity, or they may be acid-washed, typically in 5 % acetic acid solution. This also reduces their reactivity.

The glass component has to perform a number of functions in the cement. It acts as a source of ions for the initial cross-linking reaction, and also as a source of inorganic species such as silicate for the subsequent post-hardening maturation process. The glass also controls the setting rate and assists in the development of translucency, an unusual property in a cement.

A large number of glass formulations have been studied as potential cement formers. The classic glasses for these cements, developed by Wilson and his co-workers in the early days of glass-ionomers, are based on the systems SiO_2–Al_2O_3–CaO and SiO_2–Al_2O_3–CaF_2 to which other components, such as P_2O_5, Na_2O or Na_3AlF_6 are added. All of the commercially successful formulations contain fluoride. This has the advantage that it acts as a flux, thus lowering fusion temperatures. It also improves the handling characteristics of the cement paste, increases the strength of the set cement, and has the therapeutic effect when released from the set restorative, as described earlier. On the other hand, it has the disadvantage that it tends to form HF on reaction with moisture left in the glass-making ingredients, which may be lost as such from the melt or may form the volatile SiF_4. In either case, such losses result in uncontrolled alterations in the elemental composition of the glass and to etching of the heating elements of the furnace, as well as to other potential environmental damage. Formation of hydrogen fluoride is thus a major cause of batch-to-batch variations in the glass. It may be overcome either by careful drying of the reactants or by appropriate formulation of the constituents.

As was stated earlier, typical glasses for cement formation are calcium aluminosilicates. However, there have been variations on these. For example, the calcium can be replaced partly or wholly by strontium or lanthanum. Completely different glass systems have been developed, too, such as aluminoborate and zinc silicate glasses, both of which are discussed in more detail later on.

In order to react satisfactorily with a polycarboxylic acid, the initial concept was that glasses needed to be more basic than those used in the former dental silicate cement. This was achieved by having a higher alumina/silica ratio in the fusion mixture. The discovery of the rate-enhancing effect of (+)-tartaric acid by Wilson et al. [1.42] altered this formal requirement since tartaric acid is a strong enough acid to react readily with these older glasses. Consequently a satisfactory glass-ionomer cement can be made from an appropriate mixture of tartaric acid, poly(acrylic acid), water and a dental silicate glass. Nonetheless, most commercial products employ glasses of the higher alumina/silica ratio recommended by Wilson and his co-workers in their early publications on these materials [1.70–1.72].

The properties of the glasses can be understood in terms of Zacheriasen's Random Network model [1.73]. This model considers a glass to consist of a random assembly of oxygen polyhedra, each individual polyhedron comprising a central cation, such as silicon, surrounded by a small number of oxygen atoms. Thus the building block of the ionomer glasses are [SiO_4] tetrahedra. These are linked only at the corners to form chains, so that the overall glass may be viewed as a kind of highly cross-linked polymer of –Si–O–Si– units [1.74].

In the simplest case, glass would consist of an infinite network of [SiO_4] tetrahedra. It would be electrically neutral and impervious to acid attack. This situation may be altered by the introduction of so-called network-modifying cations, such as Ca^{2+}. To accommodate such an ion, the network needs to acquire a negative charge, which it does by incorporating an additional oxygen ion: Consult original printout from author for equation

$$-Si-O-Si- \xrightarrow{Ca^{2+}} -Si-O^- Ca^{2+} {}^-O-Si-$$

Aluminium included in the glass will also modify the structure of the glass, but in one of two possible ways. It may either adopt four-fold co-ordination in an oxygen polyhedron and become a network former, or it may adopt six-fold co-ordination, essentially as discrete Al^{3+} ions, and act as a network modifier. Where aluminium acts as a network former, it also exercises a modifying rôle, because it formally provides only three positive charges in the oxygen poly-hedron, as against four provided by silicon, this leaving a net negative charge on the structure. This must be balanced by the inclusion of correspondingly posi-tively charged ions to sustain electroneutrality. Aluminium is able to behave in this way due to the very similar size of the formal cations Al^{3+} and Si^{4+}, though there are limits to the extent to which it can occur. Above an aluminium to silicon ratio of 1:1, aluminium is no longer forced to adopt the preferred struc-ture of the $[SiO_4]$ tetrahedra, and hence it no longer acts as a charge-deficient network former [1.75].

The existence within the glass of non-bridging oxygen atoms, and of nega-tively charged aluminium sites, renders it susceptible to acid decomposition. The mode of attack is illustrated in Fig. 1.5.

Attack by acid involves initial attack by acidic protons at the sites of the net-work-modifying cations Ca^{2+} and Na^+. This leads to rupture of the network at

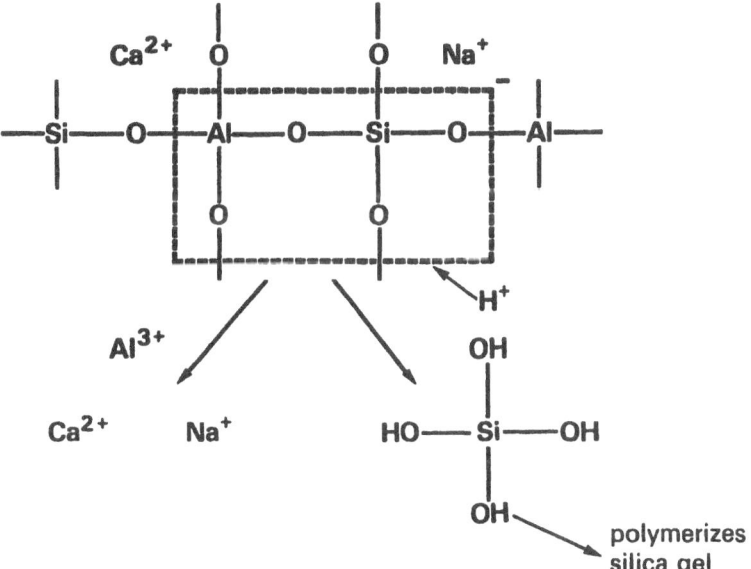

Fig. 1.5. Diagram of the setting reactions in a glass polyalkenoate cement

the aluminium sites to yield silicic acid and aluminium ions [1.5] and these species enter the cement matrix to take part in the cement hardening reactions [1.76].

The reactivity of the glass towards the polyacid depends on its basic character. This basicity has been considered in terms of the balance between acidic and basic components of the pre-firing mixture. Acidic components are oxides of silicon, boron or phosphorus, while basic ones are metal oxides, including Al_2O_3, and also calcium oxide. In purely oxide glasses, the basicity depends on the ability of the oxygen atoms to take up electrons [1.77], and is greatest for Na^+ and Ca^{2+} as cations, and least with Si^{4+} as the notional cation.

To form satisfactory cements in a clinically acceptable time, the glass has to have an appropriate basicity which, as we have seen, is governed by the SiO_2/Al_2O_3 ratio. In general, if this ratio is below 2.0, the cements will form in satisfactory times, i.e. 2.5 – 5.0 min. The SiO_2/Al_2O_3 ratio, in influencing the setting rate, alters other properties of the cement, including the strength. Opacity of the cement is also affected. All of these are illustrated in Fig. 1.6.

The glasses used in polyalkenoate cements are not necessarily homogeneous, but in most cases are phase-separated. This introduces another variable into the glass-making process, and the extent to which phase separation is allowed to occur in the quenching stage will influence the properties of the cements made from that glass. In general, phase separation reduces the reactivity of the glass, as active species become locked into the different phases. These phases may consist of discrete droplets in a matrix, or of two essentially continuous phases that effectively consist of a connected droplet-like phase in a matrix-like phase, the result of so-called spinodal decomposition.

In setting, phase-separated glasses undergo selective attack at the droplets that are rich in calcium and fluoride [1.78]. In general, the cements prepared from these glasses are stronger than those prepared from single-phase glasses. By way of illustration, the strongest cements produced from single phase glasses have compressive strengths around 130 MPa, whereas phase-separated glasses will yield cements with strengths over 200 MPa. The presence of fluoride, too, leads to stronger cements; the strongest cement produced from a pure oxide glass has a strength only just in excess of 100 MPa.

Practical glass polyalkenoate cements are made from a complex glass system that includes, in addition to silica and alumina, calcium fluoride, aluminium phosphate, cryolite (Na_3AlF_6) and aluminium fluoride. This glass system was examined in detail by Barry et al. [1.79] who found it to be heavily opal and phase-separated, with droplets rich in calcium and fluoride that had a complex morphology. These droplets represented 20% of the glass volume, and included significant amounts of fluorite. On altering the fusion temperature to 1300°C from 1150°C, fluorine was lost and the droplet size decreased. The resulting glass was more reactive towards aqueous poly(acrylic acid).

Two alternative series of glasses have also been considered for use in glass-ionomer cements, the aluminoborates of Combe et al. [1.80] and the zinc silicates of Darling and Hill [1.81]. The first of these has received rather more attention in the literature, with accounts of their preparation [1.80], cement forming properties [1.82] and glass heat-treatment to modify reactivity for

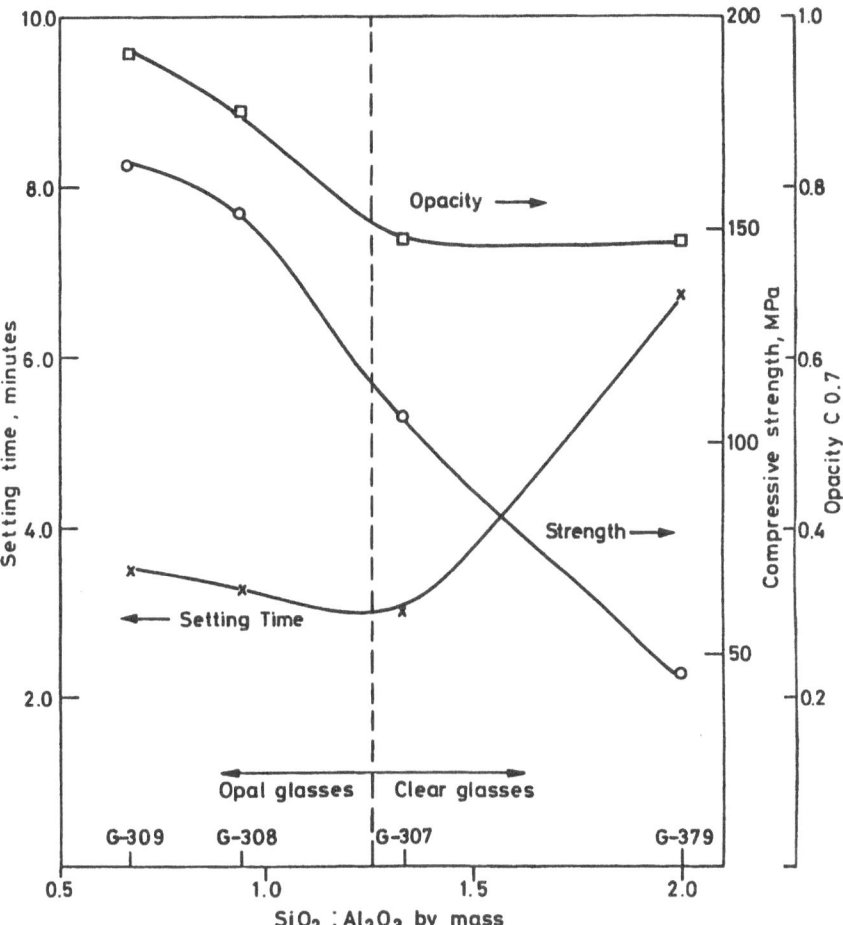

Fig. 1.6. Variation in cement properties of a glass polyalkenoate with differences in the SiO$_2$:Al$_2$O$_3$ ratio of the glass

cement preparation [1.83]. Cements made from these glasses were shown generally to have enhanced properties when 5% (+)-tartaric acid was present; setting tended to be sharper, and compressive and tensile strengths were improved. As for the conventional calcium aluminosilicate glasses, the reactivity of the aluminoborate glasses was found to vary with alumina content. In general, the higher the alumina content, the less reactive the glass towards aqueous poly(acrylic acid) [1.83].

Despite all the work that has been done on these glasses, even the best of them failed to reach the requirements of the current ISO Standard for dental restorative glass polyalkenoates. This is a serious failing, given the extent to which modern commercial glass-ionomers exceed these requirements.

A further failing of the cements made from aluminoborate glasses was that they showed relatively poor hydrolytic stability. They were, for example, signifi-

Table 1.5. Properties of some selected ZnO–SiO$_2$ and their cements with poly(acrylic acid); (after Hill and Darling [1.81])

SiO$_2$	0.57	0.42	0.42	0.42	0.46
ZnO	0.29	0.53	0.53	0.51	0.44
Al$_2$O$_3$	–	–	0.05	0.07	0.10
CaO	0.14	0.05	–	–	–
Working time/s	42	66	78	42	108
Compressive strength (MPa; 3:1 ratio, 24 h)	14.0	54.8	93.2	32.2	31.3

cantly stronger when stored dry than when stored in water [1.82], the converse of the case for cements made from conventional glasses.

The glasses prepared by Darling and Hill [1.81] were in the series CaO–ZnO–SiO$_2$ and Al$_2$O$_3$–ZnO–SiO$_2$. In both series, the ZnO and the SiO$_2$ levels were very high, and the third metal oxide was present in very low mole fraction. Selected results from this study are shown in Table 1.5, from which it can be seen that compressive strength did not begin to approach that required for commercial glass-ionomers. On the other hand, all cements were prepared at 3.0 g glass powder to 1.0 ml of 40% mol/mol poly(acrylic acid), with no attempt at optimisation, e.g. of the powder/liquid ratio or of the polymer concentration. It is likely, therefore, that better cements could be made from glasses in this series. The work was of value in demonstrating that glasses other than the usual calcium aluminosilicates may be useful in clinical glass-ionomer cements, and that materials that are effectively hybrids of glass-ionomers and zinc poly-carboxylates can be fabricated.

These glasses formed cements that did not depend on the alumina/silica ratio of the parent glass, unlike the conventional calcium aluminosilicates. Instead, the ability to form a cement was determined largely by the connectivity of the glass. One noteworthy finding was that the glass which gave the highest compressive strength (i.e. 93 MPa) has an aluminium to silicon ratio of 1:9, which for the calcium aluminosilicate glasses is too low to give a successful glass-ionomer cement. Moreover, contrary to findings for these more complex conventional glasses, increasing the mole fraction of alumina did not necessarily reduce the working time and certainly did not increase the compressive strength. Thus the empirical rules derived for the calcium aluminosilicate cements were shown not to apply to these zinc silicate glasses.

1.5.6
Polymers

The main acids used in glass-ionomer cements are polymers with carboxylic acid functional groups. The original substance used was poly(acrylic acid), a polymer still employed in many of the commercial cements. Other polymers used include copolymers of acrylic acid with maleic acid and with itaconic acid

[1.84, 1.85]. Other monomers have been used to made copolymers with acrylic acid for experimental glass-ionomer cements, such as 3-butene-1,2,3-tricarbo-xylic acid [1.5]. These polymers are all water soluble and are prepared by free-radical polymerisation in aqueous solution. The chain transfer agent employed is propan-2-ol [1.86], which is more acceptable for a clinical cement than the mercaptans used in preparing technical grade poly(acrylic acid).

The polymer is typically concentrated to the range 40–50% for use. Poly(acrylic acid) is readily soluble in water and obtaining a 50% solution presents no problems. However, at this concentration, solutions of poly(acrylic acid) tend to gel with time [1.87], a phenomenon that is attributed to the slow formation of intermolecular hydrogen bonds. Copolymers of acrylic acid with itaconic acid are more stable in aqueous solution, possibly because their assumed random nature interrupts the regularity needed to bring about gela-tion through hydrogen-bond formation.

A number of commercial glass-ionomers employ the alternative tactic of using dry acid blended with glass powder, activation of the cement being brought about by the addition of water [1.88]. There is also the possibility of dividing the poly(acrylic acid) between the powder and the liquid, and activat-ing the cement with a dilute solution of polyacid. Such cements are referred to as *semi-hydrous* [1.88].

The polymer influences strongly both the setting characteristics and the ulti-mate mechanical properties of the cement. In general, decreasing the proportion of aqueous polymer solution in the cement reduces the setting time, but in-creases the compressive strength of the set cement. The molecular weight of the polymer affects compressive strength, fracture toughness and resistance to erosion and wear, all being improved as molecular weight increases [1.90–1.92]. In addition, setting rate is increased, thus reducing working time, a feature which places an important practical limitation on the extent to which cement properties can be improved by this route. A practical upper limit on molecular weight seems to be approximately 75 000.

Most practical glass-ionomer cements make use of the rate-modifying properties of (+)-tartaric acid. This was discovered by Wilson et al. [1.42] and the effect is profound, since (+)-tartaric acid sharpens the set without decreas-ing the working time, and also leads to an increase in the compressive strength of the set cement [1.93]. In certain instances, the presence of (+)-tartaric acid allowed a cement to be formed from a glass that otherwise would not react to form one [1.94]. No other additive has proved to have quite this combination of effects, though many others have been tried [1.95]. Other multifunctional carboxylic acids, such as citric, were found to have similar effects but to raise the water solubility of the set cement to an unacceptable level.

Metal salts have been found to affect the setting of glass-ionomers. For ex-ample, divalent metal fluorides were shown to accelerate cement formation [1.95] and to increase the strength of the set cements. This effect was enhanced by the presence of (+)-tartaric acid.

More recently, Nicholson [1.96] has investigated the effect of various sodium salts on the setting and strength of an anhydrous (i.e. water-activated) glass-ionomer cement. In this study, in addition to pure water, aqueous solutions of

Table 1.6. Effect of sodium salts on the setting and strength of a glass polyalkenoate cement [1.95]

Activating liquid	Setting time/min	Compressive strength/MPa
Water	14.5	85
1 mol/l × NaCl	16.0	53
1 mol/l × NaF	22.0	80
1 mol/l × Na$_2$SO$_4$	23.0	65
1 mol/l × NaNO$_3$	17.0	65

sodium chloride, fluoride, sulphate and nitrate, all at concentrations of 1.0 mol/l were used to activate the setting reaction. Results are shown in Table 1.6.

All of the commercial glass-ionomers are polyalkenoates, i.e. are derived from carboxylic acid polymers prepared by addition polymerisation. In addition, there has been a substantial body of work published describing an experimental system based on poly(vinyl phosphonic acid). The resulting cement is properly called a polyphosphonate cement, and though covered by the trivial name glass-ionomer, this material is not included in the chemically correct title glass polyalkenoate.

Cements based on poly(vinyl phosphonic acid) (see Fig. 1.7), PVPA, have been described in papers mainly by Ellis and Wilson [1.3, 1.98, 1.99]. They are prepared from PVPA, which is itself prepared by the free radical polymerisation of the acid chloride monomer vinyl phosphonyl dichloride. This reaction employs azo-bis-isobutyronitrile as initiator, and is carried out in chloroform. The resulting polymer is then hydrolysed by treatment with water [1.100, 1.101]. PVPA is a stronger acid than poly(acrylic), and therefore requires different glasses for successful cement formation [1.102].

PVPA itself forms only indifferent cements; they are relatively weak in comparison to conventional glass-ionomer cements, and set more rapidly. However, modification of the polymer by heating with zinc fluoride or zinc phosphate reduces reactivity slightly, thereby giving a cement with a more sensible setting time, one moreover where there is sufficient time to mix and fully disperse an adequate quantity of glass. These cements have properties comparable with current commercial glass polyalkenoates.

Finally, experimental cements have also been made from blends of PVPA and poly(acrylic acid) [1.103], materials that are glass polyalkenoate/polyphosphonate hybrids. As is typical for polymer solutions, aqueous PVPA cannot be combined with aqueous poly(acrylic acid) without a two-phase system developing. In these cements, therefore, dry powders of the respective polymers were blended with appropriate amounts of glass powder and the cement activated by the addi-

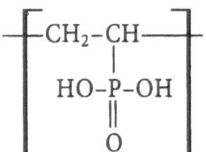

Fig. 1.7. The structure of the repeat unit in poly(vinyl phosphonic acid)

tion of water. For some of the experiments, the ZnF_2-modified variant of PVPA was employed. Optimum results were obtained for cements relatively high in PVPA, typically of the order of 9:1 of PVPA to poly(acrylic acid). However, none of the cements matched all of the properties required for glass polyalkenoates as specified in the relevant ISO Standard. One formulation, however, did approach 160 MPa which, while not quite equal to the best commercial glass-ionomers, did exceed the minimum required compressive strength of 130 MPa.

1.5.7
Structural Studies

A large number of studies have been carried out on the setting of glass-ionomer cements. Early on, infrared spectroscopy was used to demonstrate that the setting process involved the formation of calcium and aluminium polyacrylates [1.104]. This was possible because these two metals give rise to different carboxylate bands, as shown in Table 1.7. Moreover, the rate modifier (+)-tartaric acid also gives two bands, depending on whether the calcium or aluminium salt is involved, and these bands are distinct from those of the corresponding metal polyacrylates. This means that all four species can be detected in a well resolved spectrum.

On mixing the components of the cement (either aqueous polymer with glass powder or water with a mixed polymer/glass powder), hydrogen ions rapidly attack the glass particles. At the point of attack, the glass dissolves completely, releasing silicic acid and Al^{3+}, Ca^{2+}, Na^+ and F^- ions [1.106]. It is doubtful that free Al^{3+} ions exist under the conditions prevailing in setting cements. Fluoride is known to form strong complexes of the type AlF_2^+ and AlF^{2+} [1.107–1.109], species which have also been assumed to occur in these cements.

Estimates have been made of the time-dependent variation in the concentration of soluble ions in setting and hardening cements. Al^{3+}, Ca^{2+} and F^- rise to maxima as they are released from the glass, only to decrease as they become immobilised in the setting matrix [1.110]. The initial gelation of the cement is caused by formation of calcium polyacrylate, a finding that reflects the sharp rise and early fall in the concentration of soluble Ca^{2+} ions in the diagram (Fig. 1.8).

The results of the early studies on the setting of glass-ionomer cements were interpreted as implying a sequential release of ions from the glass, with calcium being released first, followed later in the setting reaction by aluminium. This was

Table 1.7. Infrared spectroscopic bands of reference salts (after Nicholson et al. [1.105])

Salt	C–O stretch of salt	
	Asymmetric	Symmetric
Ca-PAA	1550	1410
Al-PAA	1559	1460
Ca-tartrate	1595	1385
Al-tartrate	1670	1410

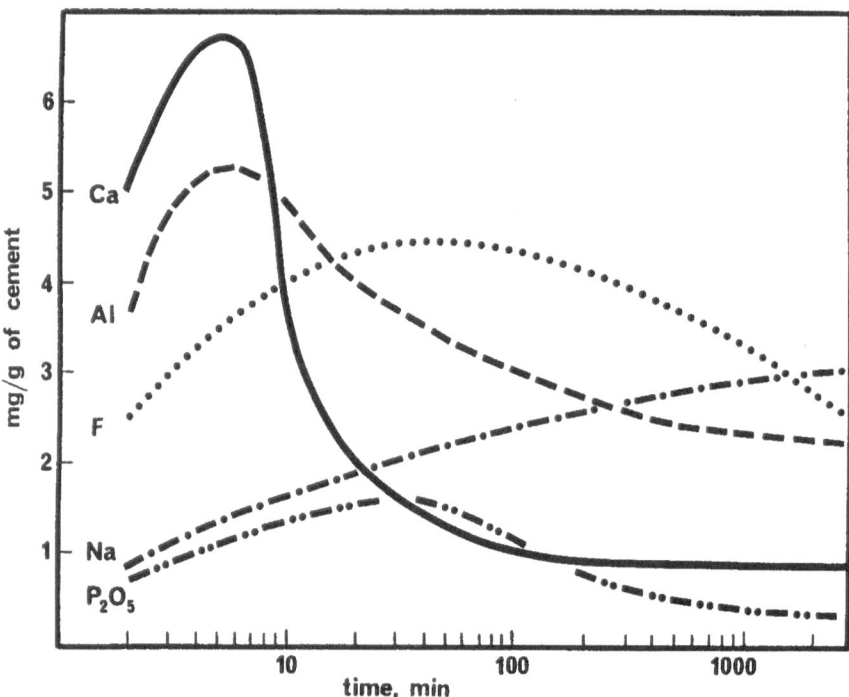

Fig. 1.8. Variation of concentration of soluble species in a setting glass polyalkenoate cement

assumed to be consistent with the results of electron probe studies, too, which showed that attack occurred preferentially at calcium-rich sites. It was unable, however, to account for the results of Cook [1.106], who examined the composition of freshly prepared cements by dissolving them in 3% sodium hydroxide solution. Atomic absorption spectroscopy of the resulting solutions revealed the presence of aluminium right at the start of the setting process, a finding which led Cook to conclude that both calcium and aluminium were implicated in the initial setting reaction [1.106].

This discrepancy led Wasson and Nicholson [1.78] to investigate further the early stages of the setting process. They did so by washing cement-forming glasses with aqueous solutions of acetic acid, following which the acid solutions were analysed by Inductively Coupled Plasma Optical Emission Spectroscopy (ICP-OES). Two glasses were used: the single phase oxide glass, known as MP4, and a complex phase-separated fluoroaluminosilicate glass, known as G338. Details of their post-firing composition are given in Table 1.8. Results from this study are shown in Tables 1.9 and 1.10.

For the glass MP4, very little change was seen in the ratios of ions either with acid concentration or with length of wash. These ratios were also, within experimental limits, those of the elements in the measured overall composition of the glass, a finding which indicates that acid attack occurs uniformly at the surface of the glass.

Table 1.8. Post-firing compsitions of glasses (used in the study by Wasson and Nicholson [1.78])

Code	Composition
MP4	Al_2O_3 (35%); SiO_2 (28%); CaO (26%); Na_2O (11%)
G338	Al_2O_3 (14.2%); SiO_2 (24.9%); CaF_2 (12.8%); AlF_3 (4.6%); $AlPO_4$ (24.2%); $Na_3 AlF_6$ (19.2%)

Table 1.9. Ion release from glass MP4

Acid strength (%)	Time (min)	No. of moles released ($\times 10^3$)				Mole ratios	
		Al	Ca	Si	Na	Al:Si	Al:Ca
2.5	60	8.75	6.24	6.08	0.79	1.40	1.44
10.0	60	17.92	14.30	12.51	1.87	1.27	1.43
5.0	2	2.93	2.17	2.01	0.29	1.35	1.46
5.0	5	5.49	3.95	3.75	0.43	1.39	1.46
5.0	10	6.50	4.64	4.54	0.62	1.40	1.43
5.0	60	12.05	8.41	8.35	1.04	1.43	1.44

Table 1.10. Ion release from glass G338

Acid strength (%)	Time (min)	No. of moles released ($\times 10^3$)				Mole ratios	
		Al	Ca	Si	Na	Al:Si	Al:Ca
2.5	60	4.04	1.95	2.75	3.07	1.47	2.07
10.0	60	10.44	3.72	6.61	6.82	1.58	2.81
5.0	2	3.52	1.62	2.16	2.37	1.62	2.17
5.0	5	4.11	1.85	2.76	3.04	1.49	2.22
5.0	10	4.12	1.81	2.86	3.06	1.44	2.28
5.0	60	6.35	2.68	4.25	4.46	1.49	2.37

For G338, there are subtle differences in the findings. As had been suggested previously by Cook, aluminium was found to be released under all conditions of acid concentration and time of washing. For G338, though, the Al:Ca ratio is smaller than the ratio of these ions present in the glass, a finding which supported the earlier work of Barry et al. [1.79] who found that acid attack occurs preferentially at calcium-rich sites.

Since aluminium was clearly released under all conditions, Wasson and Nicholson went on to explain the later formation of aluminium polyacrylate, suggesting that it was not released in a form capable of reacting with carboxylic acid groups. They suggested that a complex structure, similar to the ion $[Al_{13}O_4(OH)_{24}(H_2O)_{12}]^{7+}$, which is known to occur in aqueous solution [1.111], might be released initially. Its slow decomposition in the presence of poly (acrylic acid) would then explain the delayed formation of aluminium polyacrylate in the cement matrix.

A wider conclusion of Wasson and Nicholson concerned the rôle of the silica and alumina released from the glasses and the strength of the cement matrix. This led to a further study, in which more concentrated aqueous solutions of acetic acid were used to prepare cements in their own right [1.112]. Such cements were not instantly stable in water, but became so if allowed to age for at least 6 h. Using infrared spectroscopy, this time-scale was shown to exceed the time required for complete neutralisation by a very considerable margin. Moreover, the relevant metal acetate salts are generally extremely soluble in water, making them unlikely candidates for the rôle of matrix-formers in a water-insoluble cement. Taken together, these results led Wasson and Nicholson to postulate that the later stage of hardening of a glass-ionomer cement is formation of a silicate, or possibly phosphate, network. Previous studies, having demonstrated the gradual post-hardening maturation that manifests itself in increases in compressive strength, translucency and an increase in the ratio of bound to unbound water, were consistent with this suggestion. Moreover, the increase in proportion of bound water further suggests that this later network-forming process involves a hydration reaction.

This revised concept for the setting of glass-ionomer cements received support from a microstructural study by Hatton and Brook [1.113]. They studied thin sections of set cements in the transmission electron microscope, and found by X-ray analysis that silicon was present throughout the matrix, as well as in the glass and in ion-depleted haloes around the individual glass particles. They also showed high levels of phosphorus in the matrix, a result which was not commented upon, but which indicates there is more to learn about the details of the matrix of these cements.

Although the precise nature of the inorganic component of the matrix is not fully clear, the current hypothesis remains that it is some kind of hydrated silicate/phosphate. The extent to which it may contribute to post-hardening development of strength has also been considered [1.76]. In this paper, results are reported from experiments in which acetic acid-glass cements were aged for various periods of time and then tested for compressive strength. As before, the glass used was G338, and this was used to form cements with concentrated solutions of acetic acid ranging from 40 to 50%. Results are shown in Table 1.11.

Table 1.11. Mean compressive strengths of glass-acetic acid cements (after Wasson and Nicholson [113]). (Standard deviations in parenthese)

Storage time	Compressive strength (MPa) for					
	40%		45%		50%	
1 day	4.8	(2.4)	11.9	(3.3)	12.3	(3.4)
1 week	29.3	(8.2)	24.8	(14.8)	31.0	(14.6)
1 month	45.9	(17.2)	72.0	(27.7)	93.6	(24.9)
3 months	109.1	(22.5)	104.4	(25.2)	81.5	(28.3)
6 months	120.1	(27.3)	104.5	(25.3)	99.1	(30.1)

These results show a steady increase in strength with time that was generally significant to at least $p > 0.05$ up to 3 months. Thereafter the change from 3 months to 6 months was not significant. It was thus apparent that these cements increase in strength with age, though with greater scatter than in glass-ionomer cements, a result that perhaps indicates that the acetic acid cements are less homogeneous than properly formulated glass-ionomers.

In previous work, the increase in strength has been attributed to hydration of the polyacrylate chain [1.114, 1.115], an interpretation which cannot be applied to explain the same phenomenon in the acetic acid cements. It thus seems that the hydrated silicate/phosphate phase contributes not only further insolubility but also to the gradually increasing strength of the set cement. In contrast with the glass-ionomers, zinc polycarboxylates do not increase in strength beyond 24 h [1.29]. Consistent with this result was the finding of Wasson and Nicholson that zinc oxide and acetic acid did not form stable cements, no matter how long they were left to react before being immersed in water, and despite the rapid neutralisation undergone by the acetic acid on mixing with ZnO [1.76].

The acetic acid-glass cements were all weak at 24 h, by contrast with glass-ionomers, for which strength at 24 h can be quite considerable, for example in excess of 200 MPa. This is almost certainly due to the formation of the stiff polyacrylate matrix identified in earlier studies, which contributes to both initial strength and water resistance.

1.5.8
Biocompatibility

An important feature that has led to the great success of these materials is their biocompatibility towards both enamel and dentine [1.116]. These cements have also been shown to be biocompatible towards bone for orthopaedic applications [1.117, 1.118], as well as for restoration of the alveolar ridge in edentulous patients [1.119, 1.120]. Their good biocompatibility has also been exploited in using them in ear, nose and throat surgery, either for fixation of cochlea implants, or for sealing defects through which cerebro-spinal fluid would otherwise leak [1.121].

Glass-ionomers appear to be less biocompatible towards the tooth pulp, though there is debate about the extent to which their use is associated with pulpal irritation [1.122, 1.123]. Generally less inflammation has been found with glass-ionomer cements than with zinc phosphate or dental silicate cements [1.124, 1.125], but there remains a distinct inflammatory response which exceeds that of zinc oxide-eugenol cements [1.126]. This adverse response tends to cease within about 30 days and appears to be associated with the early stages of the setting process.

In general, more problems of pulpal sensitivity have been associated with the water-activated versions of these cements than with the hydrous ones. It has been assumed that this is because of the length of time taken for the polyacid to dissolve in the water, which would keep the pH lower for longer than in a hydrous cement [1.121]. However, studies which followed the pH change over time for both hydrous and anhydrous glass-ionomers showed that, far from the pH of anhydrous cements remaining lower for longer, it rose at rates at least equal to, and possibly slightly higher than those observed in a hydrous cement, i.e. a cement fabricated from a pre-dissolved solution of poly(acrylic acid) [1.89]. This was consistent with the finding that there was no statistically significant difference between the reported discomfort of patients treated with water-activated or with hydrous glass-ionomer cements [1.127].

Cytotoxicity of glass-ionomers has also been studied. For example, Dahl and Tronstad tested these cements against HeLa cells [1.128] and found that there was some evidence of toxicity in freshly prepared cements. By contrast, in cements that had been aged for at least 24 h, there was little or no toxic response in the HeLa cells. This correlates well with what is known about the setting chemistry of these cements, that they are sensitive to early exposure to moisture and will release a variety of water-soluble species in the first few hours after mixing.

A number of cytotoxicity studies have been reported on a particular glass-ionomer cement prepared from a copolymer of maleic and acrylic acids and undergoing development for a range of surgical applications, including reconstruction surgery following removal of bone tumours [1.129], as well as reconstruction of the alveolar ridge. Brook et al. [1.130] showed that when this material was presented as granules to cultured cells of rats or of baboons it provoked a favourable biological response. In another study, Meyer et al. [1.131] studied osteoblast biocompatibility of this material and showed it to be excellent. They also showed that aluminium was leached from the cement, and could be found in the cells, yet there was no evidence of cytotoxicity. This surprising finding they attributed to the possible formation of complexes with silica, also known to be released from the cement, in a manner previously described in the literature [1.132]. Meyer et al. also showed that bovine osteoblast cells attached themselves extremely rapidly to glass-ionomer cement surfaces compared with other biomaterials, a feature assisted by the good wettability and ionic nature of the cement surface [1.133].

An extensive study of the biocompatibility of glass-ionomers, in both bone cell cultures and in animal models, was reported by Sasanaluckit et al. [1.134]. They used a variety of materials, including commercial cements made from maleic/acrylic acid copolymer, cermet-containing cements, experimental cements based on poly(acrylic acid) and on poly(vinyl phosphonic acid), and one resin-modified liner (Brand: Vitrebond, ex. 3 M Dental Products). They determined acute cytotoxicity using contact techniques with cultured fibroblasts, then they evaluated toxicity following extraction. They also studied the effect of implanting smooth rods of set cement into the femur of adult hooded rats. Results between materials and between evaluation techniques varied quite widely, but some general conclusions could be drawn. Firstly, conventional glass-ionomers showed very good biocompatibility in each test regime, with the experimental PVPA cements showing extremely good results in the direct contact test, with significantly more bone growth than any other cement, and no zone of inhibition around them. Secondly, Vitrebond showed consistently and significantly inferior biocompatibility by all three test methods. It was, however, slightly less poor following extraction, suggesting that whatever caused the poor biocompatibility was leached out by the extraction process. No conclusions were drawn, however, concerning the identity of the toxic species.

In their major review of the topic of biocompatibility of glass-ionomers, Nicholson et al. [1.116] suggested that there were essentially three reasons for the good biocompatibility of conventional self-hardening glass-ionomers. They are

(a) the minimal setting exotherm [1.135];
(b) the rapidity of the neutralisation process which removes potentially irritating acid; and
(c) that the substances leached from the cement are generally either benign or beneficial towards the tissue in which the cement is placed. Each of these features will now be considered in detail.

The setting exotherm of acid-base cements varies considerably. In one study, Crisp et al. [1.135] found that, under controlled conditions of specimen size and using a specially designed calorimeter, glass-ionomer cement gave the smallest setting exotherm of all the cements examined. Its temperature rose a mere 3.9 °C compared with a maximum temperature rise of 22.1 °C recorded for the zinc phosphate cement. Since excessive temperature rise can damage the tooth pulp [1.136], this very low setting exotherm for the glass-ionomer cement contributes to its biocompatibility.

The second feature contributing to good biocompatibility is the speed of neutralisation following mixing. The parent polymeric acids anyway tend to be weak. Poly(acrylic acid), for instance, has a pK_a of 4.5–5.0 depending on concentration [1.137]. In a study of the progressive neutralisation of poly(acrylic acid) with sodium hydroxide, this was shown to rise to between 6 and 7.5 as neutralisation was approached, the actual pH depending on the concentration of the polymer.

A number of studies have been made of the change in pH on setting of glass-ionomers. Two techniques have been used. In one, the surface pH has been determined using a flat-headed electrode in contact with a piece of moist filter paper pressed up against a disk of setting cement. In the other, for the first part of the setting process, the pH of the paste has been measured directly; subsequently, individual specimens of cement have been crushed and extracted with a small amount of water, the pH of which has then been measured. This latter technique gave values that have been designated pH_α and pH_β respectively [1.138]. The value of pH_α or pH_β has been shown to be higher for a given glass-ionomer cement at a given time than the surface pH. For example, Wasson and Nicholson showed a particular glass-ionomer cement, Baseline (ex. DeTrey Dentsply) had a pH_β of 4.94 at 60 min from mixing [1.89] whereas Woolford [1.139] had previously shown the surface pH to be still below 4 at 60 min.

Although the precise value of measured pH varies with the technique used to determine it, all of the recorded figures are quite high and it is very unlikely that they lead directly to pulpal irritation. Certainly, Wasson and Nicholson showed that anhydrous zinc polycarboxylates had similar values of pH_α and pH_β within the first few minutes after mixing, and that for both zinc polycarboxylates and glass-ionomers, pH was well above 3.0–3.5 within 10 min. Neutralisation approached completion (i.e. pH 7) faster in zinc polycarboxylates than in glass-ionomers, as had previously been shown by both conductivity [1.140] and spectroscopic [1.34] studies, but initial neutralisation was comparable in both materials. This, coupled with the fact that other workers have shown that maintaining restorations free from microbial contamination [1.141, 1.142] leads to low or absent pulpal irritation, suggests that it is probably not simply low pH in a glass-ionomer that causes the occasional occurrence of pulpal irritation, but the result of metabolic activity of micro-organisms.

The final feature that has been held to contribute to good biocompatibility is the generally benign nature of the species leached from set cements. The actual substances leached depend on the initial composition of the cement. One of the first studies on these cements, which perforce employed an early version of the material, so-called ASPA IV, showed that silica, sodium, aluminium, phosphate and fluoride were released but, surprisingly, no calcium [1.143]. These studies employed a buffered sodium lactate/lactic acid solution (pH 4.0) at 37 °C as extracting medium. By contrast, using a different cement formulation, with water as the extracting medium, traces of calcium were found [1.144]. The main species found in this latter study, which used a conductimetric method of measurement, were silica, sodium and fluoride. There was no phosphate in the glass employed for these cements, hence none appeared in the leachates.

The species which have been found to be released from set glass-ionomers are all inorganic, and small enough to diffuse through and out of the set cement. Indeed fluoride release is usually considered to be diffusion controlled [1.145, 1.146]. By contrast, there has never been any indication of release of the polymeric component. This component is of high molecular weight, and therefore fairly immobile, so that its non-appearance among the leachates is readily explained.

Of the inorganic substances released, silica, calcium, phosphate and fluoride are all quite acceptable biologically, especially at the doses released from a small cement specimen. Silica, for example, seems to be fairly benign [1.147], as do calcium and phosphate, which are anyway the constituents of the inorganic phase of teeth and bones [1.148]. Small doses of fluoride are useful in counteracting the development of caries, an effect that seems related to the ease with which fluoride ions replace hydroxyl groups in the mineral hydroxyapatite, making a mineral phase that more effectively resists acid attack in the mouth [1.149]. Of the ions released, only aluminium seems undesirable. However, as mentioned earlier, there does not seem to be any harmful effect attributable to aluminium in the cells, and this may be due to the formation of aluminium-silica complexes in situ.

1.5.9
Reinforced Glass-Ionomers

It was Wilson and Prosser who first reported attempts to improve the strength of glass-ionomers by reinforcement [1.150]. They did so by the incorporation of metal fibres and flakes into the setting cement, an approach that was successful in improving the flexural strength. However, there were problems in that these cements showed a correspondingly poor wear resistance and a greater tendency to fracture.

Since this early work, two approaches have been employed to reinforce these cements: (i) the incorporation of powdered metal, such as silver alloy [1.151] from conventional amalgam alloy, and (ii) the fusing of the metal, typically silver, with a reactive glass powder to make a sintered ceramic-metal powder known as a *cermet*, and using this powder to form cements by reaction with a polyacid [1.152]. Cermets for glass-ionomers have also been made from gold [1.153] but this form has not been manufactured commercially.

Since the introduction of commercial reinforced glass-ionomers, a number of workers have reported results of investigations into the effect of reinforcement on compressive, flexural, diametral tensile and bond strengths of these materials. Unfortunately, results are difficult to compare because a range of specimen sizes, loading rates and storage times have been used [1.154]. A further complication is that these cements tend to be difficult to mix by hand, and the results obtained therefore depend heavily on the skill of the person carrying out the mixing [1.155].

To give an idea of the range of results obtained, Table 1.12 lists the published results, together with experimental conditions, that have been reported for four brands of reinforced glass-ionomers.

Flexural strengths have also been determined on these cements, and show a comparable range of conditions, with a similar bewildering range of results. Again, we need to bear in mind the fact that these cements are difficult to mix by hand, and that results depend to a large extent on the skill of the operator [1.155].

Table 1.12. Compressive strengths (K) for reinforced glass polyalkenoates (from Wasson [156])

Material	Type	Sample size (height × diameter)	Loading rate (mm/min)	K (MPa)	Reference
Ketac –Silver	Cermet	6 mm × 4 mm	1	153	1.158
		?	?	144.3	1.158
		6 mm × 5 mm	1	216.9	1.159
		6 mm × 4 mm	1	198.9	1.159
		10 mm × 5 mm	0.5	150	1.160
		12 mm × 6 mm	1	175.3	1.161
		12 mm × 6 mm	0.5	150	1.162
Chelon –Silver	Alloy	10 mm × 5 mm	0.5	182	1.160
		12 mm × 6 mm	0.5	129	1.160
		12 mm × 6 mm	1	136.7	1.163
Miracle Mix	Alloy	?	?	147.5	1.162
		?	?	120	1.163
		12 mm × 6 mm	1	168.7	1.161
		12 mm × 6 mm	0.5	129	1.163
		12 mm × 6 mm	1	137.6	1.164
		?	?	197	1.165
RGI-reinforced	Alloy	12 mm × 6 mm	1	221.5	1.158

A question that is raised by these results is whether the addition of either metal alloy or cermet really reinforces the resulting cement. In the case of the cermet system, there is considerable doubt that it does. Williams et al. [1.164] studied both original versions and versions of cements with additions of cermet or metal, and showed that the cermet cement was weaker than its equivalent cermet-free cement, whereas the metal-containing cement was genuinely reinforced by comparison with its unfilled analogue. Similar results were found for flexural strengths. These results have been confirmed and shown to hold regardless of the skill of the operator [1.155].

The cermet- and metal-containing cements have been used as core build-up materials, in combination with conventional glass-ionomer cement to secure crowns in place. Comparisons have been made with amalgam for this purpose [1.165, 1.166] and it has been found that the bond strength of the casting was the same regardless of core material. For the cermet- or metal-containing glass-ionomer, however, bond strength was improved by thermocycling, whereas amalgams were unaffected.

The state of the remaining tooth determines the clinical technique that should be employed, since DeWald et al. [1.166] showed that adhesion of the reinforced cement was adequate only where at least two walls of the tooth remained. Where less tooth remained, retention in the form of pins and grooves was found to be necessary.

Fluoride release has been shown to occur from most of the cermet- and metal-containing cements at levels which compare with those of conventional glass-ionomers. One exception is Ketac-Silver (ex. ESPE), the cermet-containing glass-ionomer, which has been shown to release much less fluoride and which occasionally permits slight caries development [1.167]. Thus the amount of fluoride released by this cement (1.5 ppm at 2 months, declining to 0.7 ppm at 5 years) probably represents the threshold for effective clinical release. The presence of metal alloys leads to some silver release from these cements. Sarkar et al. [1.168] have investigated the in vitro release of silver from both Ketac-Silver and Miracle Mix (ex, GC) using direct anodic stripping voltametry. They showed clear release of silver, with Miracle Mix, the alloy-containing material, releasing slightly greater amounts than the Ketac-Silver.

Clinically, cermet- and metal-containing glass-ionomers have been used for a variety of purposes. In addition to core build-up prior to placement of crowns, they have been used for Class I and Class II restorations in primary teeth. A number of studies have confirmed their effectiveness in these applications [1.169, 1.170] and their cariostatic properties in clinical use. Class II lesions in molars can be treated by the so-called tunnel approach, in which carious material is removed in the form of a tunnel, which is then overfilled with cement. When it has hardened, the excess is removed and the cement trimmed and finished. Croll, for example, has reported the successful clinical application of this technique for both primary [1.171] and permanent [1.172] molars, using Ketac-Silver as the glass-ionomer cement.

Materials other than metal or cermet powders have been used in attempts to reinforce glass-ionomers. In an early study, for example, Prosser at al. [1.173]

Table 1.13. Effect of adding hydroxyapatite to glass polyalkenoate cements. (Standard deviations in parentheses)

Glass type	Filler loading %	Working time (min)	Setting time (min)	Compressive strength (MPa)	
MP4	0	2	9	20	(1)
	2.5	3	19	20	(1)
	5.0	6	34	18	(1)
	10.0	10	55	14	(2)
	20.0	10	76	11	(1)
G200	0	2	11	55	(8)
	2.5	2	12	53	(4)
	5.0	2	13	56	(5)
	10.0	3	12	53	(7)
	20.0	3	12	52	(3)
	25.0	3	12	50	(6)
	30.0	4	20	49	(7)
	40.0	5	16	48	(3)
	50.0	12	59	26	(2)

employed metal fibres. Unfortunately, these cements were more prone to wear than the unfilled cement, hence this approach was abandoned.

Oldfield and Ellis [1.174] investigated the use of both carbon and aluminosilicate fibres to reinforce glass-ionomers, as a possible route to the development of glass-ionomer cements for use in orthopaedic surgery. This led to increased Young's modulus and increased flexural strength, but unfortunately these reinforced cements proved difficult to mix.

Finally, Nicholson et al. [1.175] studied the inclusion of finely divided hydroxyapatite into glass-ionomers. This was done because other biomaterials, including zinc polycarboxylate dental cement [1.179], had previously been filled with hydroxyapatite, with considerable enhancement of their properties, including biocompatibility for specific applications.

The effects of this inclusion were found to vary with the glass used. In the case of the single phase oxide glass MP4, there was a steady increase in the working and setting times and a steady decrease in the compressive strength (see Table 1.13). By contrast, for the phase-separated fluoride-containing glass G200, whilst there was also a steady increase in working and setting times, the compressive strength did not decrease steadily. Instead there was a plateau region from 0–40% hydroxyapatite loading in which compressive strength did not decline by a statistically significant amount.

The origin of the differences between these glasses was not clear. In the case of cements prepared from the fluoride glass, there were parallels with the behaviour of zinc polycarboxylate, which also exhibited a plateau region where increasing the amount of hydroxyapatite added did not significantly alter the properties of the set cement [1.175].

1.6
Resin-Modified Glass-Ionomers

1.6.1
Basic Chemistry

Resin-modified glass-ionomer cements for use as lining materials in dentistry became commercially available in the late 1980s [1.177]. They are hybrid materials prepared by the incorporation of photopolymerisable components into a conventional acid-base mixture [1.178, 1.179]. They consist of a complex mixture of components, and include:

(i)   poly(acrylic acid) or a modified poly(acrylic acid) altered by introducing polymerisable side chains to the polyacid molecule;
(ii) a photocurable monomer, typically hydroxyethylmethacrylate, HEMA;
(iii) possible additional photopolymerisable molecules, such as bis-GMA or similar substance;
(iv) an ion-leachable glass;
(v)   water.

These cements set by a number of competing reactions and they have complex structures. They were launched initially as liners/bases, and as such have been shown to have good adhesion to bovine dentine [1.180] and to release clinically useful amounts of fluoride [1.181]. Later, restorative grades of resin-modified glass-ionomer became available, and longer term exposure studies have shown that one of these materials, Fuji II LC (ex. GC, Japan), swelled slightly in water, though it did not become weaker, and remained brittle in fracture [1.182].

In considering the particular case of the components of resin-modified glass-ionomers, we need to be aware of the general case of the interaction of two polymers in solution. It is well established that the general result of mixing two polymers, whether in solution or in the melt, is a two-phase system [1.183]. This is explained in terms of thermodynamics:

The free energy of mixing can be calculated from the Gibbs equation:

$$\Delta G_{mix} = \Delta H_{mix} - T \Delta S_{mix}$$

For mixing to occur spontaneously a necessary (but not sufficient) condition is that ΔG_{mix} be negative. Because the number of specific polymer-polymer interactions is small, due to steric effects, enthalpy changes for mixing of solutions of polymers tend to be small [1.184]. At the very least, this makes the ΔH_{mix} term small, so that it makes only a minor contribution to ΔG_{mix}. For chemically dissimilar polymers, favourable interactions between similar segments of the same kind of molecule may be disrupted, thus making ΔH_{mix} positive, i.e. opposed to the mixing process.

Entropy of mixing, too, is small, for reasons that become clear from the application of the Flory-Huggins lattice theory. This states that:

$$\Delta S_{mix} = - R (N_1 \ln \phi_1 + N_2 \ln \phi_2)$$

where R is the gas constant, N_i is the number of moles of component i and $_i$ is the volume fraction of component i. Since polymers have a high molecular weight, the number of moles per unit part of the solution is small. Due to the small size of the N_i terms, the ΔS_{mix} term is small; in the limit of infinite molecular weight, it is zero.

The overall effect of these two thermodynamic terms being small is that ΔG_{mix} for polymers is itself small. Thus there is little thermodynamic driving force for spontaneous mixing, with the net result that blends of polymer melts or of polymer solutions are generally not miscible.

These broad generalisations do not apply in all circumstances, for instance:

(i)   for oligomers, the entropy of mixing term is not negligible;
(ii) with chemically similar polymers, where the enthalpy of mixing is likely to be small, very low ΔS_{mix} may be sufficient to drive the process; and
(iii) where polymer can develop large and specific interactions, such as hydrogen bonding, between different types of molecule; in such cases ΔH_{mix} is large and negative, and hence swamps the small ΔS_{mix} term.

The above argument is expressed in terms of thermodynamic miscibility. For real polymers, however, it may not be necessary to achieve true miscibility. A stable, intimate dispersion of one polymer solution phase in the other may be sufficient. Provided the dispersion remains intact over reasonable time periods, a condition known as *compatibility*, the mixture may be satisfactory. For the polymeric components of resin-modified glass-ionomers, while true miscibility probably cannot be achieved, with care the more limited goal of compatibility can be.

None of the commercial resin-modified glass polyalkenoates has been available for very long, hence any problems of slow phase separation may not have had the chance to occur. Thus, in the main, compatibility does seem to have been achieved. This may be helped by the fact that, in a number of cases, the components of these cements are copolymers. Copolymers are generally more miscible with other polymers than homopolymers, a fact which arises because interactions with the second polymer reduce any unfavourable interactions between the different segments of the same molecule [1.185].

The presence of organic molecules in an aqueous solution has the effect of altering the conformation of dissolved polyelectrolytes. This has been shown experimentally for poly(acrylic acid) in the presence of methanol [1.186]. Methanol was found to reduce the "goodness" of water as a solvent for the polymer, thus moving the system closer to the Flory theta-temperature, i.e. closer to phase separation. Although the actual experiments were carried out under what are formally described as *semi-dilute* conditions, hence far removed in terms of concentration from the polymer solution used in glass-ionomer cements, there are parallels between this study of poly(acrylic acid) solutions and an early study of glass-ionomer cements. Some years ago, in an attempt to overcome gelation in concentrated aqueous solutions of poly(acrylic acid), Crisp et al. [1.187] added methanol to the mixture. They found that the setting reaction of such mixtures with a calcium fluoro-aluminosilicate glass was slower than for poly(acrylic acid) in water alone, a fact they attributed to esterification of some of the carboxylic acid groups on the polymer. It seems likely, particularly in view of the findings of Klooster et al. [1.186], that the methanol had reduced the solvating power of the water, and led to the polymer adopting a more tightly coiled (hence less reactive) conformation. Similar effects seem likely to occur with HEMA in resin-modified glass-ionomer cements.

Setting chemistry is complicated for resin-modified glass-ionomers. In principle, on irradiation the setting occurs rapidly by the photochemical cross-linking reaction, and more slowly by the acid-base reaction. In practice these two reactions cannot take place without reference to each other: the photochemical reaction will be affected by the polar nature of the acid-base medium, and the acid-base process will be affected by the presence of relatively hydrophobic entities, and also by the reduced diffusion coefficients of the reactive species through the cross-linked network.

In a study designed to model what happens in a resin-modified glass-ionomer, Anstice and Nicholson used 50/50 water/methanol and water/HEMA mixtures to bring about setting in a water-activated glass-ionomer based on poly(acrylic acid) [1.188]. They showed that in each case the setting reaction was slowed down significantly and that the resulting cements were weaker than those prepared from water alone. The use of the water/HEMA combination had less effect on the speed of the setting reaction than the use of water/methanol, but it led to weaker cements (Table 1.14). Both mixtures, however, gave cements that exhibited the post-hardening maturation process typical of glass-ionomers based on poly(acrylic acid) which gradually led to increased compressive strength (Table 1.15).

Table 1.14. Effect of organic compounds on measured compressive strength (Standard deviations in parentheses)

Liquid	Storage time/days	Compressive strength/MPa		Change (% of Initial Value)			
				Volume		Mass	
Water	1	230	(34)	−0.9	(1.0)	1.0	(0.2)
	7	281	(21)	−1.6	(1.2)	−1.2	(0.3)
	30	303	(20)	0.2	(1.2)	1.6	(0.4)
Water/ MeOH	1	170	(18)	0.9	(0.9)	2.3	(0.2)
	7	204	(20)	0.2	(0.6)	2.8	(0.3)
	30	231	(13)	0.4	(0.5)	3.2	(0.2)
Water/ HEMA	1	147	(13)	4.6	(0.9)	2.2	(0.1)
	7	147	(13)	5.3	(0.7)	3.5	(0.6)
	30	168	(7)	4.8	(0.6)	3.3	(0.1)

Table 1.15. Effect of organic compounds on working and setting times (Standard deviations in parentheses)

Liquid	Time (min)			
	Working		Setting	
Water	3.6	(0.1)	9.1	(0.6)
Water/methanol	12.8	(1.3)	28.0	(2.4)
Water/HEMA	6.6	(0.4)	19.1	(0.1)

In addition, the presence of the polar poly(acrylic acid) molecule in water is likely to alter the rate of the photopolymerisation reaction, though it is not clear whether or not this is disadvantageous. Conversely, the presence of the non-polar photopolymerisable molecules (and, in certain resin-modified cements, segment) will alter the rate of the acid-base reaction, and this will be disadvantageous. This effect would compound the already reduced rate of acid-base reaction experienced with the reduction in diffusion coefficients for the bulky reacting species as the photopolymerised network undergoes development.

A further feature of the setting reaction is a natural tendency for the reacting mixture to phase-separate as the reaction proceeds. Recent experimental work has shown that phase-separation occurs when HEMA is copolymerised with ethylene glycol dimethacrylate in aqueous solution in the presence of an ionic salt [1.189]. This system is clearly a relevant model for what happens in a resin-modified glass-ionomer cement.

There are two driving forces for phase-separation. Firstly, as HEMA undergoes polymerisation, it ceases to be water-soluble [1.190]. This will compound the general tendency of mixtures of polymer solutions to separate. Secondly, as the acid-base reaction proceeds, and the poly(acrylic acid) becomes progressively more neutralised, so more hydrophobic organic species become less soluble in the aqueous phase. This is the well-known phenomenon of *salting out*, which is responsible for such effects as the development of cloudiness when salts are added to surfactant-water systems [1.191].

The phase-separating tendency as resin-modified glass-ionomers undergo setting means that the product itself is likely to contain domains of different phases. There are parallels in the microstructure of materials such as carboxylated rubbers [1.192] and lightly sulfonated fluoropolymers [1.193]. These latter materials have been particularly widely studied and shown to have a clustered morphology in which the ions, with some co-ordinated water molecules, are associated in domains in an otherwise hydrophobic fluoropolymer matrix. These domains are formed due to the highly unfavourable thermodynamic state of neutralised functional groups plus ions in a non-polar medium. Small angle X-ray scattering has been used to determine the size of these domains in certain of these materials [1.194] and there is no question that they occur in a variety of related materials. Similarly, they have been assumed to occur in ethylene-acrylic acid copolymers containing only low levels of neutralisable functional groups [1.195]. It thus seems to be a general structural feature of polymeric materials based on mixed hydrophobic/ionic components that the ions form discrete assemblies within the organic matrix. Although this kind of microstructure has not been demonstrated in fully cured resin-modified glass-ionomers, it has been shown for one restorative material that was allowed to cure by the acid-base process only [1.196]. However, as yet there is no direct experimental evidence for such hydrophobic domains in properly cured resin-modified glass-ionomers. It is also not clear what effects such domains might have on the longer term durability of these cements.

1.6.2
Structural Studies

On light activation, a resin-modified glass-ionomer cement undergoes its photochemical polymerisation, thus becoming less susceptible to water exchange. The extent of the cure depends on the depth of light penetration into the cement, and varies with colour, efficiency of the activator light and time of irradiation. Any cement into which insufficient light has penetrated is still able to undergo so-called self-hardening, i.e. neutralisation of the carboxylic acid groups by the basic glass powder. This reaction will occur at a slower rate than in the unmodified cements, but clearly takes place. The evidence for this is that maturation in the form of a gradual increase in compressive strength is known to occur over a period of at least a week in resin-modified cements.

1.6.3
Biocompatibility

Sasanaluckit et al. [1.134] studied one of the resin-modified glass-ionomers, namely Vitrebond (ex. 3 M Dental) as part of their extensive survey of biocompatibility of these materials. They found it to give a very severe cellular response, killing all of the cells in the Petri dish when exposed in a direct contact assay to mouse fibroblast L 929. After aqueous extraction, this effect was reduced but not completely eliminated. Even after two weeks extraction at elevated temperature, the material continued to stimulate a cytotoxic response, and at no stage did cells actually grow to make contact with the material. On implantation into adult hooded Lister rats, this material showed no sign of bone formation or resorption. Instead, after an 8-week period, all that could be found around the implant was fibrous tissue.

These results were not discussed in terms of the underlying chemistry of these materials [1.134], but it was stated that some sort of extractable substance was clearly the cause of this very poor biocompatibility. In view of the known chemistry of resin-modified glass-ionomers, HEMA seems the likeliest substance to be causing the problems, though the possibility of residual monomer from the reaction used to modify the polyacid cannot be entirely ruled out. Overall, though, the results obtained by Sasanaluckit et al. [1.134] demonstrate the extremely unsatisfactory decline in biocompatibility that the resin modification process has had on this particular material, an effect that is likely to be general for all the varied types of resin-modified glass-ionomer currently available.

1.6.4
Clinical Experience

Manufacturers generally claim that a 20-s exposure to light is sufficient to promote photochemical cure in these cements. However, in clinical practice, it is often difficult to place the tip of the light source close to the surface of the cement due to the presence of the metal matrix around the site of the restoration. Hence light intensity within the cement is less than optimum. To overcome this problem, Mount [1.199] recommends a minimum of 40 s irradiation time, preferably longer for larger restorations.

Resin-modified glass-ionomers have been used as liners and bases, as restoratives and as fissure sealants. Different formulations are required for different applications. For example, the original restorative grade resin-modified glass-ionomer, Fuji II LC (GC Corporation) was prepared from higher proportions of acid-base components than the luting grades [1.198]. They were thereby stronger in compression and less susceptible to either polymerisation shrinkage or post-cure swelling in aqueous media.

Liners and bases have similar, complementary uses in clinical dentistry. A liner is placed as a thin layer of neutral material to protect the underlying pulp from thermal shock, or to make up for any deficiency in the cavity wall preparation. A base, by contrast, effectively acts as a dentine substitute, and is employed to replace major regions of dentine loss. Resin-modified glass-ionomers have been widely used in both of these ways since their launch in the late 1980s. Care is needed by the clinician to ensure that there is sufficient access for the curing light, but provided this requirement is met, there seem to be no problems with the clinical procedures involving these materials.

The use of resin-modified glass-ionomers for specific clinical purposes has been described in a number of papers. Mount [1.199] has pointed out that their principal benefit is their aesthetics, and hence they have been widely used for anterior restorations where aesthetics are most important. Restorative dentistry is less and less concerned with simple repair nowadays, and increasingly concerned with the cosmetic appearance of the finished tooth. The majority of patients no longer have rampant caries, a development which essentially leaves two problems for the clinician to deal with: (i) the repair of small lesions in a state of active caries which, though not rampant, still continue to occur despite the increased extent of prevention and the widespread presence of fluoride; and (ii) the repair of large restorations placed many years ago, when the principles of extensive cavity preparations of GV Black prevailed. In either case, the patient is increasingly likely to opt for aesthetic restorations [1.199]. Resin-modified glass-ionomers have the advantage of superior aesthetics over the self-hardening variety; they are also capable of being cut back and modified immediately after placement and cure. Resin-modified glass-ionomers have greater tensile strengths than the self-hardening materials, and this leads to better retention properties.

Croll has described a series of different restorations with different brands of resin-modified glass-ionomer cement. For example, he employed the restorative grade cement Fuji II LC (GC Corporation) in the repair of a permanent molar [1.200]. The technique involved cavity preparation, followed by the application of the cement in one portion, overfilling the cavity. Prior laboratory experiments had shown Fuji II LC to cure to a depth of at least 6 mm when subjected to a beam of blue visible light for 40 s [1.200]. Hence Croll felt able to use this length of cure to activate fully the relatively large amount of cement in the cavity, rather than to use incremental filling, layer by layer, with each layer individually photocured. Anecdotal evidence reported by Croll is that over one thousand restorations of this type have been placed in his practice without any reported tooth sensitivity. Polymerisation shrinkage under these clinical conditions was imperceptible, possibly because of the relatively high proportion of the Fuji II LC material in the conventional acid and base of the self-hardening glass-ionomer [1.200]. Despite the initial success of this and other, similar restorative procedures [1.200, 1.201], Croll recognised that durability of these repairs was uncertain, since the material has been available for such a short time that wear resistance, fracture toughness and pulpal response have yet to be fully studied and long term properties established. Nonetheless, he and Killian concluded on the up-beat note that if these materials do last reliably for 5–10 years, this is likely to lead to the complete obsolescence of silver amalgam restorations in children [1.201].

Fuji II LC is not the only resin-modified glass-ionomer to have been studied in detail clinically. Croll and Killian have also used Vitremer (3 M Dental) [1.202]. This cement is described as a tri-cure system, because it contains, in addition to the acid and base of the standard glass-ionomer, two free radical initiator systems. One of these is based on a peroxide plus amine accelerator, which generates free radicals simply by mixing at room temperature. The other is based on a camphorquinone system, and generates free radicals on exposure to light at 470 nm. Thus, when cured by irradiation, following mixing and placing, there is a surface cure via photochemical initiation, and through cure by a combination of acid-base reaction and free radical polymerisation. Though light cured for optimum properties, the existence of two through-cure reactions means that this cement has sacrificed its "command set" facility.

The Vitremer "tricure" cement system is supplied with additional components [1.202]. These are: (i) a primer, consisting of a mixture of the Vitremer copolymer, HEMA, photoinitiators and ethanol; and (ii) a finishing gloss, consisting of a clear bis-GMA/triethylene glycol dimethacrylate resin, for coating the finished restoration to control water movement in and out of the cement. As with Fuji II LC, Croll and Killian found that Vitremer could be used in large cavities in a single injection, rather than with incremental layering and curing, as for example in the recommended technique for larger composite resins. In the case of Vitremer, the "dark cure" free radical reaction contributes significantly to the overall setting process, and eliminates concern about lack of deep light penetration.

1.7
Conclusions

This chapter has surveyed the range of materials in the general class of polyelectrolyte dental cements. These cements have come a long way since the first zinc polycarboxylates appeared on the market in the late 1960s, or the first sluggishly setting opaque glass-ionomers appeared in the mid-1970s. This latter group now includes an array of variations, including improved self-hardening types, metal-reinforced and resin-modified. It is increasingly important to specify closely which one of these materials is required, since properties can now be so precisely tailored to individual clinical application.

Glass-ionomer cements are employed in an extensive set of clinical operations, from lining cavities, luting crowns, through core build-up and restoration of erosion cavities and incisal edges. They are beginning to be used outside dentistry, in fields such as maxillofacial reconstruction and cementation of cochlea implants. There is no doubt that glass-ionomers provide the solution to the treatment of a variety of diseases of the body's hard tissues. As such, they will continue to be the subject of intense study for many years to come.

1.8
References

1.1. Smith DC (1968) A new dental cement British Dental Journal 125:381–384
1.2. Wilson AD (1991) Glass-ionomer cement – origins, development and future. Clinical Materials 7:275–282
1.3. Ellis J, Wilson AD (1990) Polyphosphonates: a new class of dental material. Journal of Materials Science Letters 9:1058–1060
1.4. Mandel M (1993). In: Hara M (ed) Polyelectrolytes. Marcel Dekker, New York
1.5. Wilson AD, Nicholson JW (1993) Acid-base cements; their biomedical and industrial applications. University Press, Cambridge
1.6. Kitano T, Taguchi A, Noda I, Nagasawa M (1980) Conformation of polyelectrolytes in aqueous solution. Macromolecules 13:57–63
1.7. Jacobsen A (1962) Configurational effects of binding of magnesium to polyacrylic acids. Journal of Polymer Science 57:321–336
1.8. Oosawa F (1970) Polyelectrolytes. Marcel Dekker, New York
1.9. Wasson EA (1992) The development of glass polyalkenoate cements for orthopaedic applications (PhD thesis). Brunel University, London
1.10. Gregor HP, Frederick M (1957) Titration studies of polyacrylic and polymethacrylic acids with alkali metals and quaternary ammonium bases. Journal of Polymer Science 23:451–465
1.11. Ikagawa I, Gregory HP (1957) Theory of the effect of counter ion size upon titration behaviour of polycarboxylic acids. Journal of Polymer Science 23:477–484
1.12. Gregor HP, Gold DH, Frederick M (1957) Viscometric and conductimetric titrations of polymethacrylic acids with alkali metals and quaternary ammonium bases. Journal of Polymer Science 23:467–475
1.13. Bratko D, Dolar D, Godec A, Span J (1983) Electric transport in polystyrenesulphonate solutions. Makromolekulare Chemie Rapid Communications 4:697–701
1.14. Strauss UP, Leung YP (1965) Volume change as a criterion for site binding of counterions by polyelectrolytes. Journal of the American Chemical Society 87:1476–1480
1.15. Begala AJ, Strauss UP (1972) Dilatometric studies of counterion binding by polycarboxylates. Journal of Physical Chemistry 76:254–260
1.16. Rymden R, Stilbs P (1985) Counterion self-diffusion in aqueous solutions of poly(acrylic acid) and poly(methacrylic acid). Journal of Physical Chemistry 89:2425–2428
1.17. Rymden R, Stilbs P (1985) Concentration and molecular weight dependence of counterion self-diffusion in aqueous poly(acrylic acid) solutions. Journal of Physical Chemistry 89:3502–3505
1.18. Irving H, Williams RJP (1953) The stability of transition metal complexes. Journal of the Chemical Society 3192–3210
1.19. Wilson AD, Crisp S (1977) Organolithic Macromolecular Materials, Chaps. 2 and 4. Applied Science Publishers, Barking, Essex
1.20. Ikegami A (1964) Hydration and ion binding of polyelectrolytes. Journal of the Polymer Society A 2:907–921
1.21. Ikegami A (1968) Hydration of polyacids. Biopolymers 6:431–440

1.22. Muto N, Komatsu T, Nakagawa T (1973) Counterion effect on the titration behaviour of poly(maleic acid). Bulletin of the Chemical Society of Japan 46:2711–2715
1.23. Muto N (1974) Counterion effect on the titration behaviour of poly(itaconic acid). Bulletin of the Chemical Society of Japan 47:1122–1128
1.24. Wilson AD, McLean JW (1988) Glass Ionomer Cement. Quintessence, Chicago
1.25. Yokoyama T, Hiraoko K (1979) Hydration and thermal transition of poly(acrylic acid) salts. Polymer Preprints of the American Chemical Society 20:511–513
1.26. Hückel W (1950) Structural Chemistry of Inorganic Compounds, vol 1. Elsevier, New York
1.27. Pearson RG (1963) Hard and soft acids and bases. Journal of the American Chemical Society 85:3533–3539
1.28. Prosser HJ, Wilson AD (1979) Litho-ionomer cements and their technological applications. Journal of Chemical Technology and Biotechnology 29:69–87
1.29. Nicholson JW, Hawkins SJ, Wasson EA (1993) A study of the structure of zinc polycarboxylate dental cements. Journal of Materials Science; Materials in Medicine 4:32–36
1.30. Ikegami, Imai N (1962) Precipitation of polyelectrolytes by salts. Journal of Polymer Science 56:133–152
1.31. Nicholson JW, Wasson EA, Wilson AD (1988) Thermal behaviour of films of partially neutralised poly(acrylic acid): 3; effect of calcium and magnesium ions. British Polymer Journal 20:97–101
1.32. Wall FT, Drennan JW (1951) Gelation of polyacrylic acid by divalent ions. Journal of Polymer Science 7:83–88
1.33. Greenwood NN, Earnshaw A (1984) The Chemistry of the Elements. Pergamon, Oxford
1.34. Nicholson JW, Brookman PJ, Lacy OM, Sayers GS, Wilson AD (1988) A study of the nature and formation of the zinc polycarboxylate matrix using FTIR spectroscopy. Journal of Biomedical Materials Research 22:623–631
1.35. Mehrotra RC, Bohra R (1983) Metal Carboxylates, Academic Press, London, New York
1.36. Hill RG, Labok S (1991) The influence of polyacrylic acid molecular weight on the fracture of zinc polycarboxylate cements. Journal of Materials Science 26:67–74
1.37. Paddon JM, Wilson AD (1976) Stress relaxation studies on dental cements. Journal of Dentistry 4:183–189
1.38. Akinmade AO, Hill RG (1992) The influence of cement layer thickness on the adhesive strength of polyalkenoate cements. Biomaterials 13:931–936
1.39. Akinmade AO, Nicholson JW (1995) Adhesive layer thickness of a commercial zinc polycarboxylate dental cement. Biomaterials 16:149–153
1.40. Wilson AD, Kent BE, Clinton D, Miller RP (1972) The formation and microstructure of the dental silicate cement. Journal of Materials Science 7:220–238
1.41. Wilson AD (1968) Dental silicate cements: VII: alternative liquid acid cement formers. Journal of Dental Research 47:1133–1136
1.42. Wilson AD, Crisp S, Ferner AJ (1976) Reactions in glass ionomer cements IV; effect of chelating comonomers on setting behaviour. Journal of Dental Research 55:489–495
1.43. Crisp S, Ferner AJ, Lewis BG, Wilson AD (1975) Properties of improved glass-ionomer cement formulations Journal of Dentistry 3:125–130
1.44. McLean JW, Wilson AD (1974) Fissure sealing and filling with a glass-ionomer cement. British Dental Journal 136:269–276
1.45. McLean JW, Wilson AD (1977) The clinical development of the glass-ionomer cements I; formulations and properties. Australian Dental Journal 22:31–36
1.46. McLean JW, Wilson AD (1977) The clinical development of the glass-ionomer cements II; some clinical applications, Australian Dental Journal 22:120–127
1.47. McLean JW, Wilson AD (1977) The clinical development of the glass-ionomer cements: III: the erosion lesion. Australian Dental Journal 22:190–195
1.48. McLean JW (1987) Limitations of posterior composite resins and extending their clinical use with glass ionomer cements Quintessence International 18:517–529
1.49. Mount GJ (1992) Atlas of Glass-Ionomer Cements. Martin Dunitz, London
1.50. Crisp S, Lewis BG, Wilson AD (1976) Glass-ionomer cements: chemistry of erosion. Journal of Dental Research 55:1032–1041

1.51. Forsten L (1977) Fluoride release from glass ionomer cement. Scandinavian Journal of Dental Research 85:503–504
1.52. Tay WM, Braden M (1988) Fluoride ion diffusion from polyalkenoate (glass-ionomer) cements. Biomaterials 9:454–459
1.53. Wilson AD, Groffman DM, Kuhn AT (1985) The release of fluoride and other chemical species from a glass-ionomer cement. Biomaterials 7:55–60
1.54. Forsten L, (1991) Short- and long-term fluoride release from glass ionomer based liners. Scandinavian Journal of Dental Research 99:340–342
1.55. Tyas MJ (1991) Cariostatic effect of glass ionomer cement: a five year study. Australian Dental Journal 36:236–239
1.56. Forsten L (1993) Clinical experience with glass ionomer for proximal fillings. Acta Odontologica Scandinavia 51:195–200
1.57. Retief DH, Bradley EL, Denton JC, Switzer P (1984) Enamel and cementum fluoride uptake from a glass ionomer cement. Caries Research 18:250–257
1.58. Rolla G (1977) Effects of fluoride on initiation of plaque formation. Caries Research 11:243–261
1.59. Wei SHY (1985) Clinical Uses of Fluoride. Lea and Febiger, Philadelphia
1.60. Hamilton IR (1977) The effects of fluoride on enzymic regulation of bacterial carbohydrate metabolism. Caries Research 11 (supplement 1):321–327
1.61. Tanzer JM (1989) On changing the cariogenic chemistry of coronal plaque. Journal of Dental Research 68:1576–1587
1.62. Bellamy LJ (1975) The Infrared Specoscopy of Complex Molecules. Chapman and Hall, London
1.63. Wilson AD (1982) The nature of the zinc polycarboxylate matrix. Journal of Biomedical Materials Research 16:549–557
1.64. Crisp S, Prosser HJ, Wilson AD (1976) An infrared spectroscopic study of cement formation between metal oxides and aqueous solutions of poly(acrylic acid) Journal of Materials Science 11:36–48
1.65. Prosser HJ, Richards CP, Wilson AD (1982) NMR spectroscopy of dental cements II; the rôle of tartaric acid in glass-ionomer cements. Journal of Biomedical Materials Research 16:431–445
1.66. BS EN 29917: 1994 (1994) (ISO 9917: 1991) Specification for dental water-based cements
1.67. Anstice HM, Nicholson JW, McCabe JF (1992) The effect of using layered specimens for determination of the compressive strength of glass-ionomer cements. Journal of Dental Research 71:1871–1874
1.68. Gillmore QA (1864) Practical Treatise on Limes, Hydraulic Cements and Mortars. New York
1.69. Hill RG, Wilson AD (1988) Some structural aspects of glasses used in ionomer cements. Glass Technology 29:150–188
1.70. Wllson AD, Kent BE (1979) Polycarboxylic acid-fluoroaluminosilicate glasses surgical cement. US Patent 3 814 717
1.71. Kent BE, Lewis BG, Wilson AD (1979) Glass-ionomer formulations 1; the preparation of novel fluoroaluminosilicate glasses high in fluorine. Journal of Dental Research 58:1607–1619
1.72. Wilson AD, Crisp S, Prosser HJ, Lewis BG, Merson SA (1980) Aluminosilicate glass for polyelectrolyte cements. Industrial and Engineering Chemistry Product Research and Development 19:263–270
1.73. Zacheriasen WH (1932) The atomic arrangements in glass. Journal of the American Chemical Society 54:3841–3851
1.74. Ray NH (1983) Oxide glasses as ionic polymers. In: Wilson AD, Prosser HJ (eds) Developments in Ionic Polymers, vol 1. Applied Science Publishers, Barking, Essex
1.75. Lowenstein W (1954) The distribution of aluminium in the tetrahedra of silicates and aluminates. American Mineralogist 39:92–96
1.76. Wasson EA, Nicholson JW (1993) New aspects of the setting of glass-ionomer cements. Journal of Dental Research 72:481–483

1.77. Wilson AD (1996) Acidobasicity of oxide glasses used in glass ionomer cements. Dental Materials 12:25–29
1.78. Wasson EA, Nicholson JW (1990) A study of the relationship between setting chemistry and properties of modified glass polyalkenoate cements. British Polymer Journal 23: 179–183
1.79. Barry TI, Clinton DJ, Wilson AD (1979) The structure of a glass-ionomer cement and its relationship to the setting process. Journal of Dental Research 58:1072–1079
1.80. Neve AD, Piddock V, Combe EC (1992) Development of novel dental cements I; formulation of aluminoborate glasses. Clinical Materials 9:13–20
1.81. Darling M, Hill RG (1994) Novel polyalkenoate (glass-ionomer) dental cements based on zinc silicate glasses. Biomaterials 15:289–294
1.82. Neve AD, Piddock V, Combe EC (1992) Development of novel dental cements II, cement properties. Clinical Materials 9:21–29
1.83. Neve AD, Piddock V, Combe EC (1993) The effect of glass heat treatment on the properties of a novel polyalkenoate cement. Clinical Materials 12:113–115
1.84. Crisp S, Wilson AD (1974) Polycarboxylate cements. British Patent 1 484 454
1.85. Crisp S, Wilson AD (1977) Cements comprising acrylic and itaconic acid copolymers and fluoroaluminosilicate glass powder. US Patent 4 016 124
1.86. Smith DC (1969) Improvements relating to surgical cements. British Patent 1 139 430
1.87. Crisp S, Lewis BG, Wilson AD (1975) Gelation of polyacrylic acid aqueous solutions and the measurement of viscosity. Journal of Dental Research 54:1173–1175
1.88. McLean JW, Wilson AD, Prosser HJ (1984) Development and use of water-hardening glass-ionomer luting cements. Journal of Prosthetic Dentistry 52:175–181
1.89. Wasson EA, Nicholson JW (1993) Change in pH during setting of polyelectrolyte dental cements. Journal of Dentistry 21:122–126
1.90. Wilson AD, Crisp S, Abel G (1977) Characterisation of glass-ionomer cements 4; effects of molecular weight on physical properties. Journal of Dentistry 5:117–120
1.91. Hill RG, Wilson AD, Warrens CP (1989) The influence of poly(acrylic acid) molecular weight on the fracture toughness of glass-ionomer cements. Journal of Materials Science 24:363–371
1.92. Wilson AD, Hill RG, Warrens CP, Lewis BG (1989) The influence of poly(acrylic acid) molecular weight on some properties of glass-ionomer cement. Journal of Dental Research 68:89–94
1.93. Wilson AD, Crisp S (1975) Ionomer cements. British Polymer Journal 7:279–296
1.94. Crisp S, Merson SA, Wilson AD (1980) Modification of ionomer cements by the addition of simple metal salts. Industrial and Engineering Chemistry, Product Research & Development 19:403–408
1.95. Prosser HJ, Jerome SM, Wilson AD (1982) The effect of additives on the setting properties of a glass-ionomer cement. Journal of Dental Research 61:1195–1198
1.96. Nicholson JW (1995) Studies in the setting of polyelectrolyte materials III; the effect of sodium salts on the setting and compressive strength of glass polyalkenoate and zinc polycarboxylate dental cements. Journal of Materials Science; Materials in Medicine 6:404–408
1.97. McLean JW, Nicholson JW, Wilson AD (1994) Proposed nomenclature for glass-ionomer dental cements and related materials. Quintessence International 25:587–589
1.98. Wilson AD, Ellis J (1989) Polyvinyl phosphonic acid glass-ionomer cement. British Patent Application 2 219 289 A
1.99. Ellis J, Wilson AD (1991) A study of cements formed between metal oxides and polyvinyl phosphonic acid. Polymer International 24:221–225
1.100. Ellis J, Anstice HM, Wilson AD (1991) The glass polyphosphonate cement:a novel glass-ionomer cement based on poly(vinylphosphonic acid) Clinical Materials 7:341–345
1.101. Ellis J, Wilson AD (1992) The formation and properties of metal oxide-polyvinyl phosphonic acid cements. Dental Materials 8:79–82
1.102. Akinmade AO, Nicholson JW (1994) Development of glasses for novel polyphosphonate dental cements. British Ceramic Transactions 93:85–90

1.103. Akinmade AO, Braybrook JH, Nicholson JW (1994) Glass polyalkenoate dental cements based on physical blends of poly(acrylic acid) and poly(vinylphosphonic acid). Polymer International 34:81-88

1.104. Crisp S. Pringeur MA, Wardleworth D, Wilson AD (1974) Reactions in a glass-ionomer cement II; an infrared spectroscopic study. Journal of Dental Research 53:1414-1419

1.105. Nicholson JW, Brookman PJ, Lacy OM, Wilson AD (1988) Fourier transform infrared spectroscopic study of the rôle of tartaric acid in glass ionomer cements. Journal of Dental Research 67:1450-1454

1.106. Cook WD (1983) Degradative analysis of glass-ionomer polyelectrolyte cements. Journal of Biomedical Materials Research 17:1015-1017

1.107. Connick RE, Poulsen RE (1957) Nuclear magnetic resonance studies of aluminium fluoride complexes. Journal of the American Chemical Society 79:5153-5157

1.108. O'Reilly DE (1960) NMR chemical shifts of aluminium: experimental data and variational calculation. Journal of Chemical Physics 32:1007-1012

1.109. Akitt JW, Greenwood NN, Lester GD (1971) Nuclear magnetic resonance and Raman studies of aluminium complexes formed in aqueous solutions of aluminium salts containing phosphoric acids. Journal of the Chemical Society A:2450-2457

1.110. Crisp S, Wilson AD (1974) Reactions in glass-ionomer cements III: the precipitation reaction. Journal of Dental Research 53:1420-1424

1.111. Waters DN, Henty MS (1977) Raman spectra of aqueous solutions of hydrolysed aluminium (III) salts. Journal of the Chemical Society: Dalton Transactions:243-245

1.112. Wasson EA, Nicholson JW (1991) Studies on the setting chemistry of glass-ionomer cements. Clinical Materials 7:289-293

1.113. Hatton PJ, Brook IM (1992) Characterisation of the ultrastructure of glass-ionomer (glass polyalkenoate) cement. British Dental Journal 173:275-277

1.114. Wilson AD, Paddon JM, Crisp S (1979) The hydration of dental cements. Journal of Dental Research 58:1065-1071

1.115. Wilson AD, Crisp S, Paddon JM (1981) The hydration of a glass-ionomer cement. British Polymer Journal 13:66-70

1.116. Nicholson JW, Braybrook JH, Wasson EA (1991) The biocompatibility of glass polyalkenoate (glass-ionomer) cements. Journal of Biomaterials Science:Polymer Edition 2:277-285

1.117. Jonck LM, Grobbelaar CJ, Strating H (1989) The biocompatibility of glass-ionomer cement in joint replacement: bulk testing. Clinical Materials 4:85-107

1.118. Jonck LM, Grobbelaar CJ, Strating H (1989) Biological evaluation of glass-ionomer cement (Ketac-O) as an interface material in total joint replacement. Clinical Materials 4:201-224

1.119. Brook IM, Craig GT, Lamb DJ (1991) Initial in vivo evaluation of glass ionomer cements for use as alveolar bone substitutes. Clinical Materials 7:543-547

1.120. Duvenage JG, Jonck LM, Butow KW (1993) Porous glass-ionomer for alveolar ridge augmentation. Journal of Dental Research 74:829

1.121. Ramsden RT, Herdman RCD, Lye RH (1992) Ionomeric bone cement in neuro-otological surgery. Journal of Laryngology and Otology 106:949-953

1.122. Smith DC, Ruse ND (1986) Acidity of glass-ionomer cements during setting and its relation to pulp sensistivity. Journal of the American Dental Association 112:654-657

1.123. Stanley HR (1990) Pulpal responses to ionomer cements - biological characteristics. Journal of the American Dental Association 120:25-29

1.124. Tobias RS, Browne RM, Plant CG, Ingram DV (1978) Pulpal response to a glass-ionomer cement. British Dental Journal 144:345-350

1.125. Kawahara H, Imanshi Y, Oshima H (1979) Biological evaluation of a glass-ionomer cement. Journal of Dental Research 58:1080-1086

1.126. Plant CG, Browne RM, Knibbs PJ, Britton AS, Sorhan T (1984) Pulpal effects of glass-ionomer cements. International Endodontic Journal 17:51-59

1.127. Bebermeyer RD, Berg JH (1994) Comparison of patient-percieved postcementation sensitivity with glass-ionomer and zinc phosphate cement. Quintessence International 25:209–214

1.128. Dahl BL, Tronstad L (1976) Biological tests on an experimental glass ionomer (silico-polyacrylate) cement. Journal of Oral Rehabilitation 3:19–24

1.129. Lindeque BGP, Jonck LM (1993) Ionogram – an ionomeric micro implant in bone tumour reconstruction: a clinical evaluation. Clinical Materials 14:49–56

1.130. Brook IM, Craig GT, Hatton PV, Jonck LM (1992) Bone cell interactions with a granulated glass-ionomer bone substitute material: in vivo and in vitro culture models. Biomaterials 13:721–725

1.131. Meyer U, Szulczewski DH, Barckhaus RH, Atkinson M, Jones DB (1993) Biological evaluation of an ionomeric bone cement by osteoblast cell culture. Biomaterials 14:917–924

1.132. Birchall JD, Exley C, Chappell JS, Phillips MJ (1989) Acute aluminium toxicity to fish eliminated in silicon-rich acid waters. Nature (London) 338:146–148

1.133. Meyer U, Szulczewski DH, Möller K, Heide H, Jones DB (1993) Attachment kinetics and differentiation of osteoblasts on different biomaterials. Cells and Materials 3:129–140

1.134. Sasanaluckit P, Albustany KR, Doherty PJ, Williams DF (1993) Biocompatibility of glass ionomer cements. Biomaterials 14:906–916

1.135. Crisp S, Jennings MA, Wilson AD (1978) A study of the temperature changes occuring in setting dental cements. Journal of Oral Rehabilitation 5:139–144

1.136. Paffenbarger GC, Swanay AC, Schoonover IC, Dickson G, Glasson GF (1959) An investigation of Diafil, a dental silicate cement. Journal of the American Dental Association 39:283–287

1.137. Mandel M (1983). In: Finch CA (ed) The Chemistry and Technology of Water-Soluble Polymers. Plenum, New York, pp 179–192

1.138. Kent BE, Wilson AD (1969) Dental silicate cements VIII: acid-base aspects. Journal of Dental Research 48:412–418

1.139. Woolford M (1989) The surface pH of glass-ionomer cavity lining agents. Journal of Dentistry 17:295–300

1.140. Cook WD (1982) Dental polyelectrolyte cements I: chemistry of the early stages of the setting reaction. Biomaterials 3:232–236

1.141. Tobias RS, Plant CG, Browne RM (1987) A comparative pulpal study of two dental amalgams. International Journal of Endodontics 20:8–15

1.142. Tobias RS, Browne RM, Plant CG, Williams JA (1991) Pulpal response of two semihydrous glass-ionomer luting cements. International Journal of Endodontics 24:95–107

1.143. Crisp S, Lewis BG, Wilson AD (1980) Characterisation of glass-ionomer cements. 6. a study of erosion and water absorption from neutral and acidic media. Journal of Dentistry 8:68–74

1.144. Brookman PJ, Prosser HJ, Wilson AD (1986) A sensitive conductimetric method for measuring the material initially water leached from dental cements 4: glass-ionomer cements. Journal of Dentistry 14:74–79

1.145. Davies EH, Sefton JA, Wilson AD (1993) Preliminary study of the factors affecting the fluoride release from glass-ionomer. Biomaterials 14:74–79

1.146. Forss H (1993) Release of fluoride and other elements from light-cured glass-ionomers in neutral and acidic conditions. Journal of Dental Research 72:1257–1262

1.147. Iler RK (1979) The Chemistry of Silica. John Wiley, New York

1.148. Atkinson PJ, Witt S (1985) Characteristics of bone. In: Smith DC, Williams DF (eds) Biocompatibility of Dental Materials, Vol 1. CRC Press, Boca Raton

1.149. Nicholson JW (1994) The chemistry of teeth and bones. Education in Chemistry 31:11–13

1.150. Wilson AD, Prosser HJ (1984) A survey of inorganic and polyelectrolyte cements. British Dental Journal 157:279–282

1.151. Simmons JJ (1983) The miracle mixture: glass-ionomer and alloy powder. Texas Dental Journal 100:6–12

1.152. McLean JW, Gasser O (1985) Glass cermet cements. Quintessence International 16:333–343
1.153. McLean JW (1990) Cermet cements. Journal of the American Dental Association 120:43–47
1.154. Wasson EA (1993) Reinforced glass-ionomer cements – a review. Clinial Materials 12: 181–190
1.155. Wasson EA, Nicholson JW (1994) Effect of operator skill in determining the properties of glass-ionomer cements. Clinical Materials 15:169–172
1.156. El-Mallakh B, Sarkar NK, Kamar A (1987) Does metal incorporation improve glass-ionomer properties? Journal of Dental Research 66:115
1.157. Walls AWG, Adamson J, McCabe JF, Murray JJ (1987) The properties of a glass poly-alkenoate (ionomer) cement incorporating sintered metallic particles. Dental Materials 3:113–114
1.158. Irie M, Nakai H (1988) Mechanical properties of silver-added glass-ionomer and their bond strength to human tooth. Dental Materials Journal 7:90–97
1.159. Nakajima H, Mashimoto H, Marker VA, Hanaoka K, Miyakumi S, Teranaka T, Iwamoto T, Okabe T (1980) Static and dynamic mechanical properties of glass-ionomer. Journal of Dental Research 68:273
1.160. Kerby RE, Bleiholder RF (1989) A comparison of three metal reinforced glass-ionomer cements. Journal of Dental Research 68:250
1.161. Williams JA, Billington RW (1989) Increase in compressive strength of glass-ionomer restorative materials with respect to time: a guide to their suitability for use in posterior primary dentition. Journal of Oral Rehabilitation 16:475–479
1.162. Roeder LB, Fulton RS, Powers JM (1991) Bond strength of repaired glass-ionomer core materials. American Journal of Dentistry 4:15–18
1.163. Miller D, Marker VA, Okabe T, Simmons JJ, Zardiackas DL (1984) Formulation and evaluation of dental amalgam alloy added to glass-ionomer. Journal of Dental Research 63:231
1.164. Williams JA, Billington RW, Pearson GJ (1992) The comparative strengths of commercial glass-ionomer cements with and without metal additions. British Dental Journal 172:279–282
1.165. Arcoria CJ, DeWald JP, Moody CR, Ferracane JL (1989) A comparative study of the bond strengths of amalgam and alloy-glass-ionomer cores. Journal of Oral Rehabilitation 16:301–307
1.166. DeWald JP, Arcoria CJ, Ferracane JL (1990) Evaluation of glass-cermet cores under cast crowns. Dental Materials 6:129–132
1.167. Forsten L (1994) Fluoride release of glass ionomers. In: Hunt P (ed) Glass-Ionomers: The Next Generation, International Symposia in Dentistry, PC, Philadelphia
1.168. Sarkar NK, El-Mallakh B, Graves R (1988) Silver release from metal-reinforced glass-ionomer. Dental Materials 4:103–104
1.169. Hickel R, Voss A (1990) Clinical evaluation of glass-ionomer cement and amalgam restorations in primary molars. Journal of Dentistry for Children 57:184–188
1.170. Hung TW, Richardson AS (1990) Evaluation of glass-ionomer silver cermet restorations in primary molars:one year results. Journal of the Canadian Dental Association 56:239–240
1.171. Croll TP (1988) Glass-ionomer silver cermet Class II tunnel restorations for primary molars. Journal of Dentistry for Children 55:177–182
1.172. Croll TP (1988) Glass-ionomer silver cermet bonded composite resin Class II tunnel restorations. Quintessence International 19:533–539
1.173. Prosser HJ, Powis DR, Wilson AD (1986) Glass-ionomer cement of improved flexural strength. Journal of Dental Research 65:146–148
1.174. Oldfield CWB, Ellis B (1991) Fibrous reinforcement of glass-ionomer cements. Clinical Materials 7:313–323
1.175. Nicholson JW, Hawkins S, Smith JE (1993) The incorporation of hydroxyapatite into glass polyalkenoate ("glass-ionomer") cements: a preliminary study. Journal of Materials Science; Materials in Medicine 4:418–421
1.176. Bagnall RD, Robertson WD (1984) An adhesive dental cement containing hydroxy-apatite. Journal of Dentistry 12:135–138

1.177. Wilson AD (1990) Resin modified glass-ionomer cements. International Journal of Prosthodontics 3:425–446
1.178. Antonucci JM, McKinney JE, Stansbury JW (1988) Resin modified glass-ionomer dental cement. US Patent Application 7 160 856
1.179. Mitra SB (1988) Photocurable ionomer cement systems. European Patent Application 88 312 127.9
1.180. Mitra SB (1991) Adhesion to dentin and physical properties of a light-cured glass-ionomer liner/base. Journal of Dental Research 70:72–74
1.181. Mitra SB (1991) In vitro fluoride release from a light-cured glass ionomer liner/base. Journal of Dental Research 70:75–78
1.182. Anstice HM, Nicholson JW (1992) Study of the effect of storage in different media of Fuji II LC, a light-cured restorative glass-ionomer. LGC Occasional Paper 01/92. Laboratory of the Government Chemist, Teddington, Middlesex
1.183. Walsh DJ (1989). In: Allen G, Bevington JC (eds) Comprehensive Polymer Science, Vol 2, Chap 5. Pergamon Press, Oxford
1.184. Morawitz H (1965) Macromolecules in Solution. Interscience, New York
1.185. Stockmayer WH, Moore LD, Fixman M, Epstein BN (1955) Copolymers in dilute solution I; preliminary results for styrene-methyl methacrylate. Journal of Polymer Science 16:517–530
1.186. Klooster NTM, van der Trouw F, Mandel M (1984) Solvent effects in polyelectrolyte solutions 3; spectrophotometric results with (partially) neutralised poly(acrylic acid) in methanol and general conclusions regarding these systems. Macromolecules 17:2087–2093
1.187. Crisp S, Lewis BG, Wilson AD (1975) Gelation of polyacrylic acid aqueous solutions and the measurement of viscosity. Journal of Dental Research 54:1173–1175
1.188. Anstice HM, Nicholson JW (1994) Studies in the setting of polyelectrolyte materials II; the effect of organic compounds on a glass polyalkenoate cement. Journal of Materials Science; Materials in Medicine 5:299–302
1.189. Kiremiti M, Cukurova H, Ozkar S (1993) Spectral characaterisation and thermal behaviour of crosslinked poly(hydroxyethylmethacrylate) beads prepared by suspension polymerisation. Polymer International 30:357–361
1.190. Chirila TV, Constable IJ, Crawford GJ, Vijayasekaran S, Thompson DE, Chen Y-C, Fletcher WE, Griffen BJ (1993) Poly(2-hydroxyethyl methacrylate) sponges as implant materials: in vivo and in vitro evaluation of cellular invasion. Biomaterials 14:26–38
1.191. Vold RD, Vold MJ (1983) Colloid and Interface Science. Adison-Wesley, Massachussets
1.192. Tobolsky AV, Lyons PF, Hata N (1968) Ionic clusters in high-strength carboxylic rubbers. Macromolecules 1:515–519
1.193. Eisenberg A (1971) Glass transitions in ionic polymers. Macromolecules 4:125–128
1.194. Longworth R, Vaughan DV (1968) Physical structure of ionomers. Nature (London) 218:85–87
1.195. Bonotto S, Bonner E (1968) Effect of ion valency on the bulk physical properties of salts of ethylene-acrylic acid. Macromolecules 1:510–515
1.196. Mitra SB (1993) Private communication illustrated in H.M. Anstice, Studies on light-cured dental cements, PhD thesis. Brunel University, London
1.197. Mount GJ (1994) An Atlas of Glass-Ionomer Cements, 2nd edn. Martin Dunitz, London
1.198. Anstice HM, Nicholson JW (1994) The development of modified glass-ionomer cements for dentistry Trends in Polymer Science 2:272–276
1.199. Mount GJ (1993) Clinical placement of modern glass-ionomer cements. Quintessence International 24:99–107
1.200. Croll TP (1993) Light-hardened Class I glass-ionomer-resin cement restoration of a permanent molar. Quintessence International 24:109–113
1.201. Croll TP, Killian CM (1993) Restoration of Class II carious lesions in primary molars using light-hardening glass-ionomer-resin cements. Quintessence International 24:561–569
1.202. Croll TP, Killian CM (1993) Glass-ionomer-resin restoration of primary molars with adjacent Class II carious lesions. Quintessence International 24:723–727

CHAPTER 2

Glassy Polymers

R. L. Clarke

Glassy polymers are normally classed as amorphous and brittle. However, classifying polymers is very much a time dependent concept where a short experimental time period produces brittleness but an extended time-scale can result in viscous flow. An example of the time dependent behaviour of polymers is the material known as "bouncing putty" which flows like a viscous liquid if left under its own weight for extended periods, but shatters like a glass when hit with a hammer. Temperature is another factor which determines whether a polymer is a glassy solid, an elastic rubber or a viscous liquid. Measurements at low temperatures or high frequencies show glassy responses for polymers with Young's modulus values in the range 1–10 GPa and small strains at break. At high temperatures or low frequencies the polymer may well produce a large extension without permanent deformation with a reduced modulus in the range 1–10 MPa, thus showing a rubber-like response. The polymer finally acts as a viscous liquid at even higher temperatures where permanent deformation occurs under applied stress.

At an intermediate frequency or temperature range the phenomenon known as the glass transition (T_g) appears, where the polymer possesses an intermediate modulus between glassy and rubbery behaviour. The resultant viscoelastic response is usually accompanied by the dissipation of energy which is often expressed as the loss tangent ($\tan \delta$) of the material. Loss tangent is defined as the ratio of imaginary to real moduli. Our knowledge of the mechanical characteristics of polymers is very reliant on this vitrification temperature because of its known connection with time-temperature viscoelastic behaviour and also because insight into the origins of viscoelasticity has been possible using, for example, dielectric and dynamic mechanical relaxation techniques. Glass transition behaviour is important for dental polymers, since a denture base is required to be rigid in the oral environment at temperatures approaching the boiling point of water in the case of hot drink consumption. The major unfilled glassy polymer for such usage has, for a considerable time, been poly(methyl methacrylate). There is also a need for polymers which exhibit rubber-like characteristics at room temperature through mouth temperature and above, such as methacrylates which incorporate a plasticiser to reduce their T_g to below room temperature. These materials are used as soft liners which can be attached to the rigid denture base to form a soft cushion between the soft mucosa and the rigid denture. They are especially applicable to patients suffering effects of trauma or sore mouth conditions. Both the rigid and rubber-like methacrylates are available as heat or room temperature polymerising resins for applications as outlined above, and for use as rigid room temperature curing temporary crown and bridge resins and copy dentures. Not only are these monofunctional methacrylate monomers important dental materials but their difunctional monomers also form a large and rapidly expanding field as cosmetic composite filling materials. In this situation the difunctional methacrylate forms the matrix phase of the filled resin. Difunctionally based methacrylate fissure sealants are also available to prevent caries forming in deep occlusal fissures which are difficult to keep bacteria free.

The structure and properties of unfilled and filled methacrylate/dimethacrylate systems will now be examined in detail.

2.1
Unfilled Resins

Basically these consist of the methacrylate family, in particular methyl methacrylate and to a lesser extent ethyl, *n*-butyl and isobutyl methacrylates which form their respective polymers by free-radical addition polymerisation.

2.1.1
Monomer Preparation

An inexpensive commercial method of synthesising methyl methacrylate monomer first became available in 1932 due to J. W. C. Crawford [2.1]. Acetone was reacted with hydrogen cyanide to produce acetone cyanohydrin:

$$H_3C{\diagdown}C{=}O \quad + \quad HCN \quad \longrightarrow \quad H_3C-\underset{\underset{CN}{|}}{\overset{\overset{CH_3}{|}}{C}}-OH \tag{2.1}$$

Concentrated sulphuric acid reacted with the cyanohydrin to yield methacrylamide sulphate:

$$H_3C-\underset{\underset{CN}{|}}{\overset{\overset{CH_3}{|}}{C}}-OH \quad + \quad H_2SO_4 \quad \longrightarrow \quad H_2C{=}\underset{\underset{O}{\diagup}{\overset{C}{\diagdown}}NH_2\cdot H_2SO_4}{\overset{\overset{CH_3}{|}}{C}} \tag{2.2}$$

Esterification with methanol followed:

$$H_2C{=}\underset{\underset{O}{\diagup}{\overset{C}{\diagdown}}NH_2\cdot H_2SO_4}{\overset{\overset{CH_3}{|}}{C}} \quad + \quad CH_3OH \quad \longrightarrow \quad H_2C{=}\underset{\underset{COOCH_3}{|}}{\overset{\overset{CH_3}{|}}{C}} \quad + \quad NH_4HSO_4 \tag{2.3}$$

On purification the monomer possessed a boiling point at 760 mm of Hg of 100.5 °C, a density in the range 936–940 kg m^{-3}, and a refractive index of 1.413–1.416.

Higher alkyl methacrylates can be synthesised from methyl methacrylate by using the requisite alcohol:

$$H_2C=\underset{\underset{\displaystyle COOCH_3}{|}}{\overset{\overset{\displaystyle CH_3}{|}}{C}} + ROH \longrightarrow H_2C=\underset{\underset{\displaystyle COOR}{|}}{\overset{\overset{\displaystyle CH_3}{|}}{C}} + CH_3OH \qquad (2.4)$$

where R is the alkyl group.

2.1.2
Free Radical Polymerisation

The principal stages in the polymerisation of methyl methacrylate are the production of free radicals, known as the initiation stage, the propagation reaction and finally termination at which point the free radicals are eliminated.

Organic peroxides or azo compounds are frequently used as initiators, dibenzoyl peroxide being the common dental initiator. One way of breaking down the peroxy bond is by thermal excitation:

$$\text{(benzoyl peroxide)} \xrightarrow{60°C} 2 \text{(benzoyloxy radical)} \qquad (2.5)$$

The asterisk represents the unpaired electron. Initiation can also be achieved by:

$$\text{(benzoyloxy radical)} \longrightarrow \text{(phenyl radical)} + CO_2 \qquad (2.6)$$

In the case of room temperature polymerising methacrylates, the initiator is decomposed by incorporation of an activator in the form of a tertiary amine like N,N-dimethyl-p-toluidine [2.2]:

$$H_3C-\text{(ring)}-N''(CH_3)_2 + \text{(benzoyl peroxide)}$$

$$\longrightarrow \left[H_3C-\text{(ring)}-\underset{\underset{\displaystyle CH_3}{|}}{\overset{\overset{\displaystyle CH_3}{|}}{N}}-O-\overset{\overset{\displaystyle O}{||}}{C}-\text{(ring)} \right]^+ \text{(benzoate)}$$

$$\longrightarrow H_3C-\text{(ring)}-N^{\cdot+}(CH_3)_2 + \text{(benzoate)} + \text{(benzoyloxy radical)} \qquad (2.7)$$

According to Margerison and East [2.3], not all radicals will act as initiators and breakdown can occur as follows:

$$C_6H_5\text{-}\overset{O}{\overset{\|}{C}}\text{-}O^\bullet \ + \ C_6H_5\text{-}\overset{O}{\overset{\|}{C}}\text{-}O\text{-}O\text{-}\overset{O}{\overset{\|}{C}}\text{-}C_6H_5$$

$$\longrightarrow \ C_6H_5\text{-}\overset{O}{\overset{\|}{C}}\text{-}O\text{-}C_6H_5 \ + \ C_6H_5\text{-}\overset{O}{\overset{\|}{C}}\text{-}O^\bullet \ + \ CO_2 \tag{2.8}$$

The propagation stage uses the free radicals obtained at the initiation stage to react with methyl methacrylate. If $R^\bullet$ now represents the free radical, propagation can proceed in one of two ways:

$$R^\bullet \ + \ H_2C=\underset{COOCH_3}{\overset{CH_3}{C}} \longrightarrow R\text{-}CH_2\text{-}\underset{COOCH_3}{\overset{CH_3}{C^\bullet}} \tag{2.9}$$

or

$$R^\bullet \ + \ H_2C=\underset{COOCH_3}{\overset{CH_3}{C}} \longrightarrow R\text{-}\underset{COOCH_3}{\overset{CH_3}{C}}\text{-}CH_2^\bullet \tag{2.10}$$

Equation (2.9) is most prevalent because the $R^\bullet$ is sterically hindered by the $COOCH_3$ group in Eq. (2.10). Therefore propagation generally continues under rapid and exothermic conditions such that:

$$R\text{-}CH_2\text{-}\underset{COOCH_3}{\overset{CH_3}{C^\bullet}} \ + \ H_2C=\underset{COOCH_3}{\overset{CH_3}{C}} \longrightarrow R\text{-}CH_2\text{-}\underset{COOCH_3}{\overset{CH_3}{C}}\text{-}CH_2\text{-}\underset{COOCH_3}{\overset{CH_3}{C^\bullet}} \tag{2.11}$$

This is the common head-to-tail arrangement of molecules but as can be seen from Eq. (2.10) a head-to-head arrangement is possible:

$$R\text{-}CH_2\text{-}\underset{CH_3OOC}{\overset{H_3C}{C}}\text{-}\underset{COOCH_3}{\overset{CH_3}{C}}\text{-}CH_2^\bullet \tag{2.12}$$

Termination can occur if two growing chains meet and their radicals combine to form a stable covalent bond such as in a head-to-tail formation:

$$R(M)_xCH_2\text{-}\underset{COOCH_3}{\overset{CH_3}{C^\bullet}} \ + \ {^\bullet}\underset{COOCH_3}{\overset{CH_3}{C}}\text{-}CH_2(M)_yR \longrightarrow R(M)_xCH_2\text{-}\underset{CH_3OOC}{\overset{H_3C}{C}}\text{-}\underset{COOCH_3}{\overset{CH_3}{C}}\text{-}CH_2(M)_yR \tag{2.13}$$

$(M)_x$ and $(M)_y$ represent x and y (numbers of) repeat monomer units, respectively. Disproportionation by direct transfer of a hydrogen atom from one radical to another is also possible:

$$R(M)_xCH_2-\overset{\overset{\displaystyle CH_3}{|}}{\underset{\underset{\displaystyle COOCH_3}{|}}{\overset{\cdot}{C}}} \quad + \quad R(M)_yCH_2-\overset{\overset{\displaystyle CH_3}{|}}{\underset{\underset{\displaystyle COOCH_3}{|}}{\overset{\cdot}{C}}} \tag{2.14}$$

$$\longrightarrow R(M)_xCH=\overset{\overset{\displaystyle CH_3}{|}}{\underset{\underset{\displaystyle COOCH_3}{|}}{C}} \quad + \quad R(M)_yCH_2-\overset{\overset{\displaystyle CH_3}{|}}{\underset{\underset{\displaystyle COOCH_3}{|}}{CH}}$$

Methyl methacrylate monomers can polymerise under the influence of heat or ultraviolet light. Monomers therefore include a small quantity of an inhibitor, such as hydroquinone, which prevents spontaneous polymerisation on storage. Stable radicals are formed followed by disproportionation [2.4]. If the monomer starts to polymerise:

$$R(M)_xCH=\overset{\overset{\displaystyle CH_3}{|}}{\underset{\underset{\displaystyle COOCH_3}{|}}{C}} \longrightarrow R(M)_xCH_2-\overset{\overset{\displaystyle CH_3}{|}}{\underset{\underset{\displaystyle COOCH_3}{|}}{\overset{\cdot}{C}}}$$

Abstraction of a hydrogen atom from hydroquinone stabilises the methacrylate radical:

$$R(M)_xCH_2-\overset{\overset{\displaystyle CH_3}{|}}{\underset{\underset{\displaystyle COOCH_3}{|}}{C}}-CH_2-\overset{\overset{\displaystyle CH_3}{|}}{\underset{\underset{\displaystyle COOCH_3}{|}}{\overset{\cdot}{C}}} \quad + \quad \text{(hydroquinone, OH/OH)}$$

$$\longrightarrow \text{(phenoxy radical O}\cdot\text{/OH)} \quad + \quad R(M)_xCH_2-\overset{\overset{\displaystyle CH_3}{|}}{\underset{\underset{\displaystyle COOCH_3}{|}}{CH}} \longrightarrow \text{(OH/OH)} \quad + \quad \text{(quinone)} \tag{2.15}$$

The rate of polymerisation is decreased when oxygen is available. This is why the surface of a set room temperature cured dental acrylic or composite resin is often tacky unless air is excluded on polymerisation. Although the oxygen radical has low reactivity, it can produce oxygen-containing polymer with methacrylates but their degree of polymerisation is very low [2.5]. Radicals can also be stabilised by transfer agents that are capable of producing new radicals which in turn may instigate a fresh polymer chain.

On conversion of methyl methacrylate to poly(methyl methacrylate) at approximately 20 % conversion, the rate accelerates. At the same time the average molecular weight increases even when the reaction temperature rise is minimal. This effect is due to the chain termination reaction slowing down as the overall rate and molecular weight increase. As the reaction becomes more viscous, the ends of the polymer chains cannot diffuse towards one another, resulting in mutual termination. However, small monomer molecules can diffuse at moderate conversion with propagation reactions almost unimpaired until the mixture becomes a semi-solid at which stage the propagation rate constant decreases. This rate factor is often known as the gel or auto-acceleration effect. It is most apparent in polymers that are insoluble in their monomers, as is true for poly(methyl methacrylate) in the thermodynamic sense. Termination reactions are difficult as the radical chain ends become entrapped in the polymer. Termination then depends on the rate at which the radical ends become uncoiled at the surface where they are available for mutual termination.

2.1.3
Polymerisation by the Dough Technique

Commercially, polymerisation is by bulk or suspension techniques but in dentistry a special dough technique has been developed in which the polymerised beads are added to the monomer to form a saturated mix. This ratio is normally 3:1 by volume or 2.5:1 by weight of polymer to monomer. The volume shrinkage of methyl methacrylate on polymerisation is 21%. Using the dough method thus reduces this volume contraction to 7% since only approximately one third of the mix is monomer. On initial mixing a consistency like wet sand is formed, followed after a short while by the stringy stage when thread-like beads of polymer, which adhere to the spatula, are produced. When the polymer particles lose their tackiness and no longer stick to the mixing vessel, the dough or gelation stage has been reached. This is the condition in which the polymer is shaped into gypsum-based moulds under pressure. Beyond this stage the polymer becomes tough and rubbery, finally resulting in a reasonably hard resin. The physical chain from sandy, stringy to dough and beyond to rubbery and hard is due to the monomer penetrating the outer surface of large polymer beads which causes them to swell. At the same time small beads are completely dissolved in monomer, causing the liquid content to increase in viscosity with time. The monomer still remains unpolymerised during these stages. Heat is then required to decompose the initiator and polymerise the monomer. Alternatively, room temperature curing or autopolymerising types rely on the includ-

Table 2.1. Composition of Poly(methyl methacrylate) dough systems

Powder	Monomer
Poly(methyl methacrylate) or copolymer beads	Methyl methacrylate
Initator e.g. Benzoyl peroxide	Inhibitor e.g. Hydroquinone
Pigments	Cross-linking agent e.g. EGDM
Fibres	Plasticiser e.g. Dibutyl phthalate Activator e.g. DMPT (only in autopolymerising systems)

ed activator, such as *N,N*-dimethyl-*p*-toluidine (DMPT), to breakdown the initiator. Table 2.1 indicates the composition of these two poly(methyl methacrylate) dough systems. The powder principally contains the polymer beads, manufactured by emulsion polymerisation of the monomer, 35–200 µm in diameter. Small proportions of ethyl, butyl or other alkyl methacrylates may copolymerise with poly(methyl methacrylate) to create a more fracture resistant product. Other manufacturers may add small quantities ($\leq 5\%$) of ethyl acrylate to increase solubility and to improve flow properties in the molten state. Peroxide initiator (0.5–1.5%) is added although some is already present from the polymerisation of the beads. A cross-linking agent such as ethylene glycol dimethacrylate (EGDM) is essential to improve hardness and wear resistance of the final product.

Pigmentation to provide tissue-like colouring has been added using cadmium and mercuric sulphides, ferric oxide, carbon black and other compounds but today organic oxides are more common. Cadmium salts in particular have shown toxic tendencies although McCabe et al. [2.6] consider them to be systemically harmless. Normally the pigments are added after bead formation by ball milling with the beads. This results in an uneven colour distribution which provides a lifelike appearance to the resultant denture base. Acrylic or Nylon fibres can be incorporated to imitate the small blood vessels which underlay the oral mucosa.

2.1.4
Polymerisation by Microwaves

A recently advocated method for curing acrylic resins by the dough technique, is to use microwaves as the heat source. This requires electromagnetically generated waves in the megahertz frequency range. Typically, 2450 MHz produces a wavelength of 120 mm. In this situation methyl methacrylate molecules orientate themselves in the direction of the vibrating electromagnetic field, resulting in directional changes almost five billion times a second, according to De Clerck [2.7]. Heat is rapidly generated within the monomer due to the numerous intermolecular collisions. As the degree of polymerisation increases, the monomer content decreases proportionally, but the same quantity of energy is absorbed by less monomer as the reaction proceeds, thus increasing the activity of the remaining monomer. Theoretically this could result in complete polymerisation.

Conventional curing systems rely on monomer molecules being excited by other molecules as the consequence of thermal shock from an external heat input, and thus the products always possess a degree of residual monomer. Because microwaves are reflected by metallic objects, fibre reinforced plastic flasks must be employed and the use of desiccated gypsum for the mould is recommended to minimise water content and volume. Generally, microwave polymerised resins provide a short curing cycle (~ 3 min), have similar physical properties to conventional cures, but possess low residual monomer content and greater dimensional stability due to the excellent temperature control of the resin obtained as a consequence of their very low thermal inertia.

2.1.5
Polymerisation by Visible Light

Although the visible light cured method is not an unfilled system in that some microfine silica filler is present, it is classified as such because of its usage as a denture base material. These resins also contain a urethane dimethacrylate matrix with an acrylic based copolymer resin phase and a photoinitiator system. The pliable premixed sheets can be moulded to the master cast and polymerised by blue light of 400 – 500 nm from high intensity quartz halogen bulbs.

2.1.6
Physical Properties of Poly(methyl methacrylate)

Acrylic polymers represent a more than satisfactory match to the oral soft tissues they are replacing when considering appearance, since a subtle variety of shades plus fibre incorporation can mimic tissue of any race.

Although the temperature of the oral environment normally only ranges from 32 to 37 °C, this can increase to between – 5 and + 70 °C when the patient drinks cold and hot fluids, respectively. To ensure no distortion of the denture base occurs, the glass transition temperature must be above this range and, furthermore, a temperature exceeding 100 °C is desirable as patients have been known to soak their dentures overnight in boiling water.

Poly(methyl methacrylate) with a T_g of approximately 125 °C, for the heat cured type is thus acceptable but it must be understood that the polymer starts to soften below the transition temperature, resulting in reduced modulus which may lead to creep and distortion. Auto-polymerising acrylics are much more susceptible due to their reduced T_g of around 90 °C.

The low specific gravity (1.18) of the polymer is advantageous in the upper denture since the effects of gravitational forces which act to displace the base are minimised.

Unlike metals, acrylics are poor conductors of heat, with low values of thermal conductivity and diffusivity. This is a disadvantage as the oral mucosa beneath the denture base can become partially desensitised to the natural reflex action to hot or cold stimuli, resulting in possible patient injury. The property which controls conductor efficiency is the coefficient of thermal conductivity (k) which is the quantity of heat passing per unit area per second per unit temperature gradient. However, thermal diffusivity (D) is a far more pertinent property since it embraces the time for which the external stimulus takes to pass through the denture base and is therefore dependent on the base thickness. By definition thermal diffusivity is given by

$$D = \frac{k}{\text{specific heat} \times \text{density}}$$

Table 2.2 presents general values for the physical properties of acrylic denture base polymers discussed in this section.

The atomic constituents of poly(methyl methacrylate) are carbon, hydrogen and oxygen which provide poor radiopacity. A denture base which is detectable

Table 2.2. Physical properties of poly(methyl methacrylate) denture base

Property	Value
Specific Gravity	1.19
Glass Transition (°C) at 1 Hz	125
Thermal Conductivity ($Wm^{-1} K^{-1}$)	0.20
Thermal Diffusivity ($mm^2 s^{-1}$)	0.12

by X-ray examination is necessary if the patient swallows it or fragments of it, for example in a road accident. The problem of the radiolucent acrylic base has never been solved although numerous attempts have been made, normally failing due to aesthetics and/or mechanical properties being impaired. Use of various metal inserts gave radiopacity but inferior mechanical properties, with the possibility of digesting a fragment which was metal free [2.8]. Greatly reduced aesthetics were reported by Sainsbury [2.9] when incorporating finely divided metals such as gold flake or powdered amalgam evenly throughout the base. The majority of inorganic salts are not compatible with poly(methyl methacrylate) although barium sulphate has been employed commercially at a level of 8 wt%. Above this concentration the properties of the acrylic are severely and adversely affected. Furthermore, effective radiopacity only becomes viable

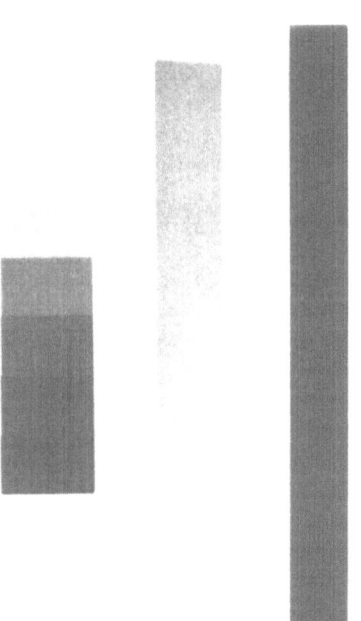

Fig. 2.1. X-ray opacity of copolymer containing poly(2,3-dibromopropyl methacrylate)

at levels of 20% barium sulphate or more. Davy and Causton [2.10] came to the conclusion that the most effective method of halide incorporation in acrylics was either to use a halogen-containing monomer or to post-polymerise pendant groups by halogenation. Their findings showed a copolymer containing 36–40% of poly (2,3-dibromopropyl methacrylate) had adequate physical properties coupled with good X-ray opacity. Figure 2.1 shows how the X-ray opacity of the copolymer increased with the percentage 2,3 dibromopropyl methacrylate in comparison with a now deleted commercial X-ray opaque denture base (Stellon). Further developments in this area have been highlighted by Davy [2.11].

2.1.7
Mechanical Properties of Poly(methyl methacrylate)

Table 2.3 lists the mechanical properties of heat cured acrylic denture bases. These indicate that the acrylic is weak, flexible and soft. However, providing the base is of adequate thickness, both strength and rigidity are generally not the cause of failure. In the oral cavity the denture base must withstand the various forces of mastication which tend to dislodge the denture. Flexural strength, which is dependent on the square of the denture base thickness, can be measured using three-point bending tests that are more representative of the loads encountered in the mouth than simple tensile and compressive tests. Momentarily high masticatory loads are normally not the cause of failure, but the poor fatigue properties of acrylic polymers are. Fatigue may become more apparent if the denture is ill-fitting and poorly designed. Another major concern with acrylics is their weak impact strength. Both these latter two inadequacies of acrylic bases will be discussed further under Modified Acrylic Polymers (Sect. 2.1.9). Lambrecht and Kydd [2.12] showed that mastication and swallowing resulted in either an increase or a decrease in curvature of the base at the mid-line, indicating the need to improve the flexural strength of poly(methyl methacrylate) in a lateral direction.

The softness of the acrylic resin, as indicated by its Vickers hardness number (Table 2.3), may suggest wear is a major concern when chewing food or cleaning the base with proprietary dentifrices. Although both the acrylic teeth and the base do show considerable attrition over long time periods, their replacement prior to such a situation arising is advisable due to ill-fit and possible repeated fracture.

Table 2.3. Mechanical properties of poly(methyl methacrylate) denture base

Property	Value
Young's Modulus (GPa)	2.6
Flexural Strength (MPa)	90
Tensile Strength (MPa)	55
Fatigue Life (cycles) at 37 °C	479000
Hardness (VHN)	20
Fracture Toughness ($MNm^{-3/2}$)	2.53

Crazing, which is first apparent as minute surface cracks, will have a weakening effect on the denture base. One known cause is the practice of some patients of allowing their dentures to dry out on removal from the mouth. The tensile stresses developed at the surface on continuous wetting and drying of the base result in crazing. Denture whitening is occasionally encountered, which is due to a mismatch of refractive indices between the bead and matrix phase caused by a structural change as a result of, for example, excessive overnight soaking temperature or exposure to solvents which can occur in saliva.

The viscoelastic nature of poly(methyl methacrylate) is manifest on subjecting the base to a constant load. The resultant strain, as a function of time, exhibits both primary and secondary creep after an instantaneous elastic response. An increase in the rate of secondary creep occurs as temperature, stress, residual monomer, cross-linking and plasticiser content increase. At low stresses, Ruyter and Espevik [2.13] found similar creep rates for heat and autopolymerised acrylics but with higher stresses the self cured resins showed greater creep rates.

2.1.8
Chemical and Biological Properties of Poly(methyl methacrylate)

All acrylic denture bases contain a small quantity of free monomer, which can in time leach into the mouth and cause irritation or more serious problems. This is one reason why autopolymerising acrylics with a higher free methyl methacrylate content than the heat polymerised variety are not considered as permanent denture bases. Presence of monomer vapour can affect the patients respiration and cardiac function, and reduce blood pressure. In its liquid form the monomer dries skin oils and can penetrate into the bloodstream at the site of a wound. Technicians are well advised to wear protective clothing and gloves when handling such inhospitable fluids.

Dimensional stability of bases is another important property which, in the case of acrylics, is mainly controlled by water absorption [2.14]. The theoretical aspects of sorption and desorption of water by polymers is, according to Crank and Park [2.15], a diffusion process which, for a flat specimen, can be regarded as unidirectional. By application of Fick's law the partial differential equation for the diffusion process is shown to be

$$\frac{\partial c}{\partial t} = D \frac{\partial^2 c}{\partial x^2} \tag{2.16}$$

where D is the diffusion coefficient and c is the instantaneous concentration at a distance x and time t. On solving this equation Crank [2.16] has shown that when t is small

$$\frac{M_t}{M_\infty} = 2 \left(\frac{Dt}{\pi l^2} \right)^{1/2} \tag{2.17}$$

where $2l$ is the sample thickness and M_t/M_∞ is the ratio of mass at time t to mass at t = ∞. This predicts the initial water uptake is linear, and by plotting M_t/M_∞

Table 2.4. Diffusion coefficient ($\times 10^{-6}\,\mathrm{mm^2\,s^{-1}}$) for water uptake in denture bases at 37.4 °C

Material	Sorption	Desorption	Equilibrium Uptake
Conventional acrylic	1.59	3.5	0.0230
High impact acrylic	1.51	3.4	0.0200
Vinyl acrylic	1.99	3.6	0.0065
Polycarbonate	6.05	10.3	0.0042

against $t^{1/2}$ the diffusion coefficient can be calculated. Thus water uptake results provide two distinct parameters according to Braden [2.17]. First, the thermo-dynamic parameter c_o, known as the equilibrium uptake, which was found to be slightly temperature dependent and, second, the diffusion coefficient which is rate dependent. This latter parameter is highly temperature dependent, approximately fitting an Arrhenius type equation [2.15]

$$D = D_o \exp\left(- E/RT\right) \qquad (2.18)$$

where E is the activation energy, R is the universal gas constant, T the absolute temperature and D_o a constant related to the entropy of activation. For poly (methyl methacrylate) Braden [2.17] calculated D to be 12 kcal/mol (50 kJ/mol). Table 2.4 itemises some typical examples of sorption, desorption and equilibrium uptake. No change in structure was evident for denture base polymers when subjected to repeated sorption/desorption cycling. The fitting surface of the denture may be colonised by organisms such as *Candida albicans* which thrive on the warm, moist conditions of the mouth. However, these are not a problem if the denture is regularly cleaned.

2.1.9
Modified Acrylic Polymers

The inherent properties of poly(methyl methacrylate) bases make them susceptible to both impact and fatigue failure as documented in Sect. 2.1.7. Consider impact failure in the first instance. Significant improvement in impact strength of glassy polymers can be achieved by the incorporation of elastomers without a corresponding decrease in modulus and heat distortion temperature (i.e. T_g), as would be expected with plasticisers. As far back as 1927, Ostromislensky [2.18] patented a method to produce toughened polystyrene by polymerising a solution of rubber in the styrene monomer. The commercial adaptation of the technique was later explored in the 1940s when the price of styrene monomer became competitive. Molecular relaxation processes have been used by Turley [2.19] to explain the effects of polymer structure on impact behaviour. He concluded that since physical properties of thermoplastics are very temperature dependent, the resultant impact strength relies on how close the polymer is to its transition temperature when tested. Near the T_g the polymer would be expected to possess a high impact strength due to its ductile/rubbery nature. In contrast, when the temperature of test is far removed from the T_g the polymer is brittle and impact strength falls. Turley also observed that many polymers with a secondary transition, due to the motion of side chains or small segments of the main chain, had good impact properties providing this relaxation was well below the test temperature. Such an example is the formerly used polycarbonate denture base with a glass transition at 150 °C, while its secondary transition, which is associated with carbonate group motion in the main chain, appears at – 95 °C. Creating an artificially low relaxation temperature for polymers like poly(methyl methacrylate) is possible by the addition of a rubber phase, which is generally a solution polymer of the rubber and methyl methacrylate. The resultant graft significantly improves impact properties. However, it is doubtful that complete grafting can occur so the system consists of ungrafted rubber, a glass-on-rubber graft and a pure glassy component. Rodford [2.20] used a rubber with reactive end groups to graft onto poly(methyl methacrylate), which had the advantage of providing a smaller rubbery domain size, thus assisting in retention of an adequate modulus.

Theoretically it has been suggested that reinforcement is concerned with the rubber phase absorbing the impact energy by mechanical damping (Buchdahl and Nielsen [2.21]). However, this does not explain the mechanism of large strain deformation. Merz et al [2.22] concluded that the rubber particles held together the opposite faces of a propagating crack, the absorbed energy on impact being the sum of the fracture energy of the glassy matrix plus the work required to fracture the rubber particles. Again this theory was incomplete, being unable to explain the difference in behaviour between certain polymers on fracture. Schmitt and Keskkula [2.23] proposed that the rubber particles did not stop crack growth but actually produced numerous energy absorbing microcracks in which context the elastomer particles acted as stress concentrators. A similar explanation was put forward by Bucknall and Smith [2.24] but microcrazes were substituted for microcracks. It was Kambour [2.25] who showed that crazing preceded fracture in glassy polymers, producing high fracture surface energies as recorded by Berry [2.26] for polystyrene and poly(methyl methacrylate). Thus the Bucknall and Smith theory was modified such that rubber particles initiated craze formation but then stopped growth into crack formation. Other hypotheses have included the dilation theory [2.27] where the rubber particles create a hydrostatic tension in the adjacent matrix, producing dilation which results in yielding in preference to brittle fracture. Bucknall [2.28] recognised that crazing and shear yielding occurred simultaneously in most rubber toughened plastics. He cited the case of high impact polystyrene where crazing is the main process without shear yielding, as opposed to acrylonitrile-butadiene-styrene where both crazing and shear yielding occur together, resulting in stress whitening and necking. The currently accepted mechanism of rubber reinforcement considers crazing to be the dominant factor, but shear yielding is evident, especially in the more ductile polymers like poly(vinyl chloride).

The number of denture repairs carried out by the National Health Service in the United Kingdom is evidence of the poor impact properties of poly(methyl methacrylate). Table 2.5 lists impact strength for various types of denture bases using a modified Charpy test, the Hounsfield. It can be observed that an increase

Table 2.5. Impact strength (Houndsfield) of acrylic denture bases

Material	Impact Strength (J)
Conventional	0.028
Autopolymerising	0.027
High Impact	0.075
Microwave	0.032
Visible Light	0.027

in value occurs for rubber modified acrylics, whereas the microwave types are similar to standard poly(methyl methacrylate) denture base. The degree of conversion of the monomer on polymerisation and the resultant final free monomer content of the autopolymerising resins is indicative of their reduced impact strength. Rodford [2.20] described the use of low molecular weight butadiene-styrene rubbers to reinforce poly(methyl methacrylate) denture bases which had a styrene content of 30 wt% and a number average molecular weight of 29900. These macromolecular monomers possessed reactive acrylate end groups which allowed grafting to the acrylic base. The author then showed that the reactive end groups cross-linked the polymer, thus improving impact strength without reducing modulus values. The macromolecular monomer was in some respects behaving like an antiplasticiser to maintain the stiffness of the resin.

The incorporation of fibres into the acrylic base has provided a method of improving flexural strength and reducing the number of mid-line fractures occurring mainly in the vulnerable upper plate, resulting in enhanced fatigue resistance and rigidity. However, at the present time their use is limited due to the increased production time involved. It is paramount that the fibres are correctly positioned in that part of the appliance which is under tensile forces in the mouth if maximum benefits are to be extracted. Another concern is that intimate bonding between the fibres and the resin base may not be achieved, resulting in a weakening effect.

Carbon fibres can be considered as the first successful reinforcement, providing increased rigidity and flexural strength. These advantages were off-set by their intense dark colour which cannot be masked, their inability to bond to the poly(methyl methacrylate) without soaking the fibres in excess methyl methacrylate monomer and the increased thickness of the base needed to accommodate multilayer, multidirectional fibres. Silane based coupling agents aid adhesion but such chemicals are potential toxins. Their biocompatibility with soft tissues, if the fibres become exposed on the surface of the denture, have been questioned by Bowman et al. [2.29]. Schreiber [2.30] demonstrated an increase in both impact strength and transverse strength with the addition of surface-treated carbon fibres to poly(methyl methacrylate). Fatigue resistance improved from 42 to 100 % by incorporation of randomly orientated carbon fibres coated with a silane coupling agent [2.31]. Carbon fibres arranged perpendicular to the direction of the applied stress proved to be the most effective in increasing resistance to bending and flexural fatigue [2.32]. Yazdanie and Mahood [2.33] found that carbon fibre strands were more effective strengtheners than a woven mat.

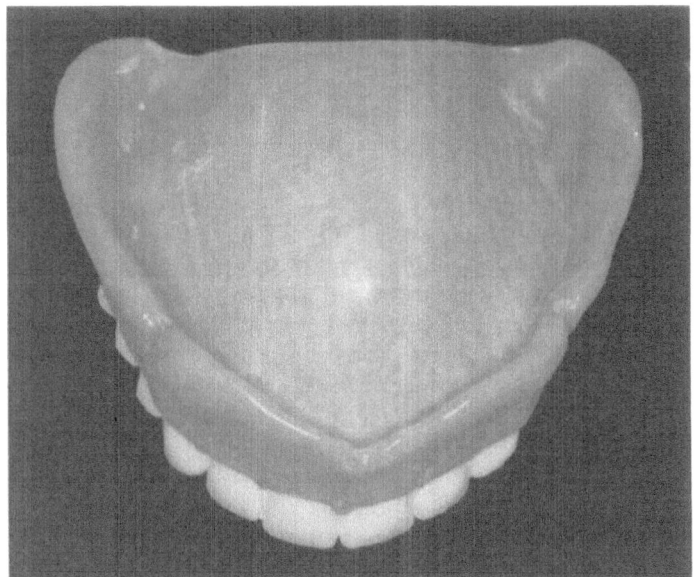

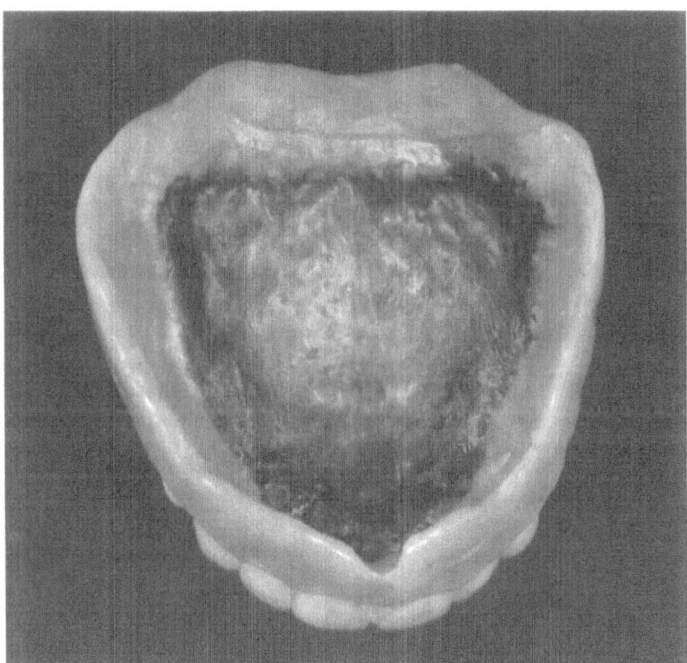

Fig. 2.2. HDLPE and carbon fibre reinforced denture bases

Woven mat is, however, the most frequently used format because of its convenience in handling during the packing stages.

A far more acceptable reinforcement, especially from the patient's viewpoint, are the aesthetically pleasing ultra high modulus polyethylene fibres (UHMPE) which are sometimes referred to as highly drawn linear polyethylene fibres (HDLPE). Polyethylene is a crystalline polymer which may be drawn, at temperatures below its melting point, into fibres which show considerable improvement in strength and modulus in the draw direction [2.34]. Such fibres are suitable for denture base reinforcement because of their properties of known biocompatibility, chemical inertness, solvent resistance, hydrophobicity and low density. Unfortunately, polyethylene possesses a low surface energy, resulting in poor wettability. This property has been improved by etching the fibres by the process of plasma treatment, which creates mechanical bonding with the matrix phase [2.35]. Reinforcement of the acrylic base by these fibres was initially performed by Braden et al. [2.36] in 1988 where they claimed considerable improvement in impact strength which was unaffected by prolonged exposure to water. Although the modulus and flexural strength of the basic acrylic are enhanced, this is not so pronounced as with carbon fibres. Recent developments in fibre reinforcement have been reviewed by Ladizesky and Ward [2.37]. Unlike Braden et al. [2.36], who used monofilaments and prewoven fibre mat, Gutteridge [2.38] incorporated 6 mm lengths of randomly orientated fibres into commercial denture bases and showed how impact strength improved with increased fibre content from 0.5 to 3 wt%. Arrangement of the fibres in a plane parallel to the surface of the denture base was found to be the most effective reinforcement. This can be achieved without the need to increase the thickness of the base providing a woven mesh of continuous fibres is incorporated. The positioning and maintenance of the fibre mesh in the correct location during processing remains a problem which, together with the additional stages involved, makes the process time-consuming. Figure 2.2 indicates the aesthetic advantage of the HDLPE fibres in comparison to the carbon fibres.

A further reinforcement known as polyaramid in the form of a plain fabric has been used with some success. Polyaramids are aromatic polyamides where the CH_2 groups in the main polymer chain of the nylon are predominately replaced by aromatic rings:

$$\sim\!\sim\!\sim NH-(CH_2)-NH-\underset{O}{\overset{\parallel}{C}}-(CH_2)-\underset{O}{\overset{\parallel}{C}}\!\sim\!\sim\!\sim \qquad \text{polyamide}$$

$$\sim\!\sim\!\sim NH-\!\!\langle\bigcirc\rangle\!\!-NH-\underset{O}{\overset{\parallel}{C}}-\!\!\langle\bigcirc\rangle\!\!-\underset{O}{\overset{\parallel}{C}}\!\sim\!\sim\!\sim \qquad \text{polyaramide}$$

Commercial production of these polyaramid fibres, poly-(p-phenylene terephthalamide), marketed as Kevlar in 1973 by the Du Pont Company, was carried out by the reaction of a diamine with a diacid chloride. These fibres are strong under tension but somewhat weak under compression. In the form of a fibre fabric they have excellent creep and fatigue resistance but their amber shade does modify the appearance of the pink denture base.

2.1.10
Higher Methacrylates

Because of the adverse effects of methyl methacrylate on pulp and soft tissue, a number of materials are available that use poly(ethyl methacrylate) with *n* or isobutyl methacrylate monomer in room temperature polymerising resins for denture relines, and temporary crowns and bridges [2.39]. Heterocyclic methacrylates have also been described [2.40]. All these systems have been shown to he biocompatible with dental tissues [2.41].

2.2
Filled Resins

The ease of fabrication and aesthetic properties of poly(methyl methacrylate) resins naturally led to their attempted use as anterior filling materials, especially once the autopolymerising technique had been perfected. Despite these advantages, their continued lack of abrasion resistance, discoloration, shrinkage (resulting in microleakage) and exothermic temperature rise on curing in the mouth has resulted in their general demise. Direct filling acrylics are still available from the major dental manufacturers but their application is restricted to temporary anterior fillings. As a first step to improve the abrasion resistance and shrinkage of such materials, glass beads were incorporated within the poly (methyl methacrylate) powder to act as reinforcing fillers, but methyl methacrylate remained the major constituent of the monomer phase with all its inherent disadvantages. Finally, alternative systems employing polyfunctional monomers and inorganic fillers began to emerge in the early 1970s, which were the forerunners of the composite filling materials of the present era.

Use of filled resins in dentistry is chiefly restricted to composite filling materials and fissure sealants which employ difunctional methacrylates as their monomer phase. Of the monomers available, the most frequently encountered are Bowen's resin and urethane dimethacrylates. However, due to the high viscosity of these monomers, diluents are added to allow them to flow sufficiently at mouth temperature for packing into the tooth cavity. Examples of diluents are

Table 2.6. Constituents of composite resins

1 Resin matrix phase	(i) Bis GMA + diluents or (ii) Urethane dimethacrylate + diluents
2 Filler phase	(i) Macro e.g. quartz, glasses (ii) Micro e.g. fumed(pyrolytic) silica
3 Coupling agent	Silane
4 Curing system	(i) Chemical e.g. benzoyl peroxide + amine or (ii) Photo e.g. Visible or UV
5 Other additives	Inhibitors, UV stabilisers, Optical brighteners, Pigments

triethylene glycol dimethacrylate (TEGDM) and 1,3-butanediol dimethacrylate. Polymerisation of the resin can be achieved by using a peroxide/amine system, as described under unfilled resins, or now more commonly by an α-diketone and an amine which are activated by the incidence of visible light. Inhibitors such as the methyl ether of hydroquinone are present in minute quantities to prevent polymerisation on storage. The filler component can be quartz, lithium aluminosilicate, barium or strontium glasses or tetragonal zirconium, which are all surface treated with a silane coupling agent to provide a chemical bond at the resin/filler interface. These macrofillers are generally combined with micro-fillers of pyrolytic silica (fumed silica) to improve the polishability and surface finish of the composite resin, especially in the anterior situation. The essential constituents of composite filling materials are summarised in Table 2.6.

2.2.1
The Monomer Phase

The resin matrix component binds the ingredients together but at the same time is responsible for the inherent shrinkage and exotherm produced on curing. These difunctional monomers have been produced by the following methods

i) Bowen's Resin; Bowen's initial work on dimethacrylate resins was the start-ing point of modern dental composite development. The first successful di-methacrylate matrix resulted from his synthesis of bis-GMA in 1962 [2.42]. Until then, methyl methacrylate was the monomer of choice. Glycidyl methacrylate was reacted with bis-phenol A such that:

$$(2.19)$$

bis-GMA: 2, 2-bis-4-(2-hydroxy-3-methacryloyl-oxypropoxy) phenyl propane

Bis-glycidyl methacrylate (bis-GMA) or Bowen's resin is a thermosetting epoxy resin with methacrylate end groups which can be cured like an autopoly-merising acrylic when suitably activated chemically or by light waves in the ultraviolet or visible region of the spectrum. A modified version, synthesised by reacting diglycidyl ether of bis-phenol A with a methacrylic acid, appeared in 1965 [2.43]. It possesses several advantages over methyl methacrylate, such as a molecular weight of 512 compared with 100 for MMA, a larger molecule which diffuses less readily, a lower thermal expansion and shrinkage and lower vola-tility. Coupled with these advantages is the problem of high viscosity caused by the bulky phenyl rings in its backbone chains and the considerable hydrogen bonding between molecules. The inability to work with such a high viscosity material at room temperature (1200 Pas) has resulted in the addition of lower viscosity monomers like TEGDM. Reduced stiffness and increased polymerisation shrinkage of the composite resin result as undesirable consequences.

Ferracane [2.44] suggested that the properties of the unfilled resin are dependent on their molecular structure and the degree of conversion of the carbon-carbon double bonds in the methacrylate groups, but that the degree of cross-linking may have an even greater effect. The degree of conversion of bis-GMA monomer has been shown to be approximately 55% for a chemically cured system by Cowperthwaite et al. [2.45] and slightly less (48%) using UV radiation [2.46]. Use of a diluent can raise the maximum conversion to more than 70% for these systems. Ferracane [2.44] indicated there was a significant correlation between diluent concentration and degree of conversion since greater molecular mobility exists with increased diluent concentration during curing. This improved degree of conversion produced stronger bis-GMA/TEGDM resins. However, the correlation was only applicable when comparing similar types of resin. In general, greater conversion does lead to improved stability of the resin and an increased degree of cross-linking above room temperature. This increased cross-link density maintains a higher level of dynamic modulus for the resin in comparison to a low cross-link density material at elevated temperatures, although in both cases modulus does fall with increase in temperature. Such a situation is to be expected since increased cross-link density indicates greater stiffness due to the reduced mobility of the polymer chains.

ii) Modified Bowen Resins; to overcome the disadvantages of using diluents with the original bis-GMA monomers, alternative systems with reduced viscosities have been developed. Of these, the hydroxy group elimination has resulted in the reduction of hydrolytic degradation. Braden and Davy [2.47] showed the absence of the OH groups lowered the viscosity of the monomer and reduced the water absorption of the cured polymer. Other versions of hydroxy free bis-GMA resins involved the absence of the propoxy group (bis-MA), use of substituted alkoxy such as ethoxylated (bis-EMA) [2.48] and bis-PMA. These monomers are illustrated in Fig. 2.3.

iii) Aliphatic dimethacrylates; the urethane dimethacrylate monomers developed in 1974 by Foster and Walker [2.49] via the reaction of an aliphatic di-isocyanate with a hydroxylalkyl methacrylate have proved less viscous than bis-GMA resins. Since no phenyl groups exist in the polymer chain their flexibility and toughness is considerably increased compared to Bowen's resin.

$$H_2C=C-C-O(CH_2)_2-O-C-NH-CH_2-C-CH_2-CH-(CH_2)_2-NH-C-O(CH_2)_2-O-C-C=CH_2 \tag{2.20}$$

UDMA: 1,6-bis-(methacryloxy-2-ethoxy carbonylamino)-2,4,4-trimethylhexane

The first commercial visible light cured dental composite (Fotofil by ICI/Johnson & Johnson) contained EGDM monomer with various UDMA isomers

(a)

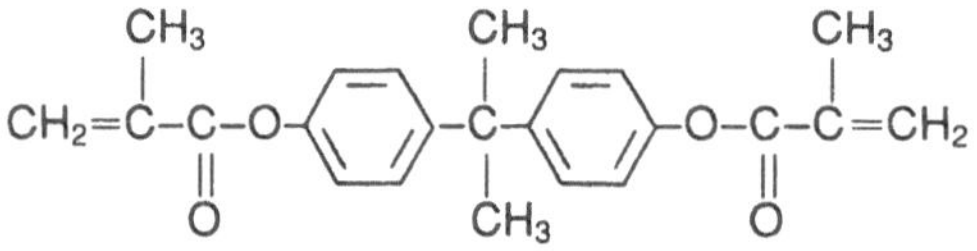

$$CH_2=\overset{CH_3}{\underset{\underset{O}{\|}}{C}}-C-O-\bigcirc-\overset{CH_3}{\underset{\underset{CH_3}{|}}{C}}-\bigcirc-O-\overset{CH_3}{\underset{\underset{O}{\|}}{C}}-C=CH_2$$

Bis-MA: 2,2-bis-4-(3-methacryloyl) phenyl propane

(b)

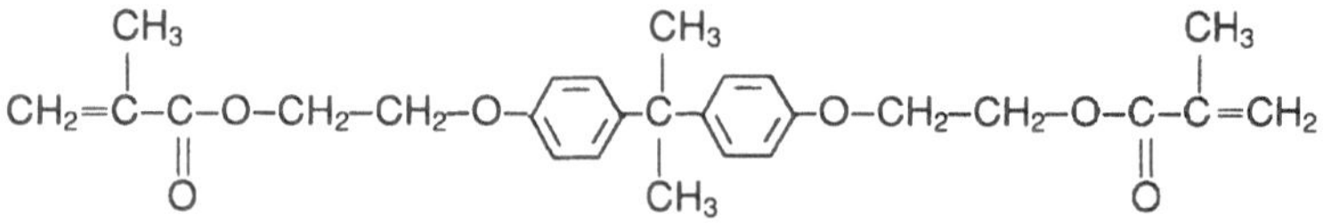

$$CH_2=\overset{CH_3}{\underset{\underset{O}{\|}}{C}}-C-O-CH_2-CH_2-O-\bigcirc-\overset{CH_3}{\underset{\underset{CH_3}{|}}{C}}-\bigcirc-O-CH_2-CH_2-O-\overset{CH_3}{\underset{\underset{O}{\|}}{C}}-C=CH_2$$

Bis-EMA: 2,2-bis-4-(3-methacryloyl-oxyethoxy) phenyl propane

(c)

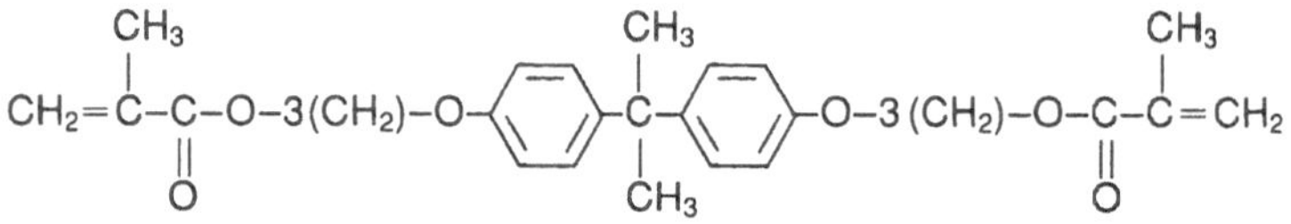

$$CH_2=\overset{CH_3}{\underset{\underset{O}{\|}}{C}}-C-O-3(CH_2)-O-\bigcirc-\overset{CH_3}{\underset{\underset{CH_3}{|}}{C}}-\bigcirc-O-3(CH_2)-O-\overset{CH_3}{\underset{\underset{O}{\|}}{C}}-C=CH_2$$

Bis-PMA: 2,2-bis-4-(3-methacryl-oxypropoxy) phenyl propane

(d)

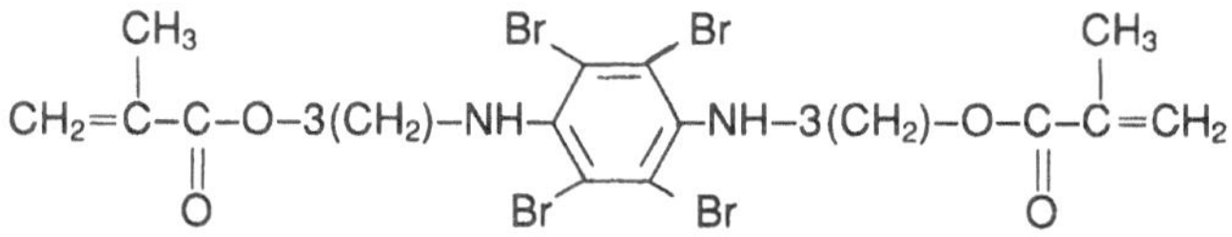

$$CH_2=\overset{CH_3}{\underset{\underset{O}{\|}}{C}}-C-O-3(CH_2)-NH-\overset{Br\quad Br}{\underset{Br\quad Br}{\bigcirc}}-NH-3(CH_2)-O-\overset{CH_3}{\underset{\underset{O}{\|}}{C}}-C=CH_2$$

1,4-Bis-(methacryloxy-N-propyl)-2,3,5,6-(tetrabromo phenylene) diamine

(e)

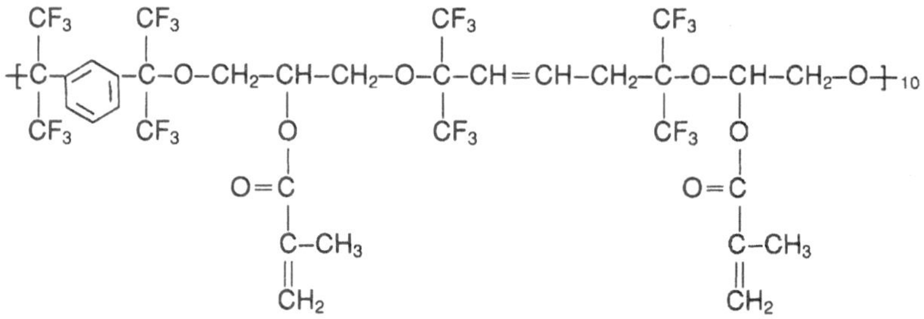

$$+\overset{CF_3}{\underset{\underset{CF_3}{|}}{C}}-\bigcirc-\overset{CF_3}{\underset{\underset{CF_3}{|}}{C}}-O-CH_2-\underset{\underset{\underset{\underset{CH_2}{\|}}{C-CH_3}}{O=C}}{CH}-CH_2-O-\overset{CF_3}{\underset{\underset{CF_3}{|}}{C}}-CH=CH-CH_2-\overset{CF_3}{\underset{\underset{CF_3}{|}}{C}}-O-\underset{\underset{\underset{\underset{CH_2}{\|}}{C-CH_3}}{O=C}}{CH}-CH_2-O+_{10}$$

Polyfluorinated polymethacrylate

Fig. 2.3 a–e. Composite monomers

and derivatives [2.50]. A simplified representation can be schematically shown as:

$$\left[\begin{array}{c} H_3C \\ H_3C \end{array} \!\!> C \!-\!\!\! \bigcirc \!\!-\! O \!-\!\! \{ CH_2\!-\!\underset{CH_3}{CH}\!-\!O \}_{\overline{mn}}\!\!\overset{O}{\underset{}{C}}\!-\!\overset{H}{N}\!-\!\!\bigcirc\!\!-\!CH_2\!-\!\!\bigcirc\!\!-\!\overset{H}{N}\!-\!\overset{O}{\underset{}{C}}\!-\!O\!-\!(CH_2)_2\!-\!O\!-\!\overset{CH_3}{\underset{O}{C}}\!-\!\overset{CH_3}{\underset{}{C}}\!=\!CH_2 \right]_2 \quad (2.21)$$

The di(urethanophenyl) methane group is unchanged when the monomer polymerises but quinoid groups are formed leading to photo-oxidation [2.51]. Exposure to artificial sunlight and water has resulted in severe yellowing of such systems. UV stabilisers were added to "Fotofil" to minimise the discoloration. More recently the sensitive di(urethanophenyl) methane group has been replaced by hexamethylene diurethane in several visible activated materials.

iv) Recent modified monomers; Arroyo Dental Products [2.52] patented a room temperature polymerisable resin in 1978 which claimed to be radiopaque. It contained a brominated aromatic diacrylate with acrylate reactive diluent (Fig. 2.3 d). Other researchers suggested the use of polyfluorinated polymethacrylate provided high strength coupled with reduced water absorption [2.53] (Fig. 2.3 e).

Ring opening systems which expand on polymerisation have been suggested to alleviate one of the major shortcomings of dental composite resins, namely their polymerisation shrinkage. Spiroorthocarbonate (SOC), when polymerised to a polycarbonate with methylene pendant groups, showed a volume expansion of 17 vol.% [2.54]. Addition of 15–20% SOC, in solid form, to bis-GMA/TEGDM resins has been investigated. Stansbury [2.55] and Millich et al. [2.56] produced rearranged structures incorporating SOC while Byerley et al. [2.57] cured such systems via cationic photoinitiation, which proved more compatible with epoxy resins. One expandable SOC studied by Eick et al. [2.58] was *trans/trans*-2,3,8,9-di(tetramethylene)-1,5,7,11-tetraoxospiro[5.5]-undecane with (4-octyloxyphenyl) phenyliodonium hexafluoroantimonate initiator and 2-chloro-thioxanthone as the sensitiser. The 5 wt% of the *trans/trans* SOC in epoxy comonomers when cured at room temperature was found to have acceptable properties and a 0.1% expansion based on a density determination. Thus non-shrinking resins may be possible but at the present time they are far from ideal with respect to optical and aesthetic characteristics.

2.2.2
The Filler Phase

Fillers are necessary for various reasons, such as improving strength, handling properties and radiopacity, reducing coefficient of thermal expansion and, by no means least, for minimising polymerisation shrinkage. A single filler may be unable to provide such a wide variety of improvements but several filler types may. The fillers themselves must in general be resistant to the chemical environment in the mouth, be colourless, non-toxic, match the refractive index of the polymer matrix, be relatively hard and have a reinforcing effect on the matrix phase. Of the few substances which can fulfil these criteria, glass and glass ceramic together with some silicates and silicone dioxide are available. Silicone dioxide is a major component of all these fillers, whose surfaces can bond to the matrix phase via silanol groups on the filler surface. The coupling agents required to create a bond between the silanol groups and the resin phase are silanes, which provide the requisite reinforcing effect of the filler particles on the composite resin.

Addition of inorganic filler particles to dental resins first became viable once Bowen had developed his now famous bis-GMA copolymer matrix system [2.42, 2.43], together with an effective coupling agent between the filler and matrix components. These early composite resins contained E-glass fibres, glass beads, soda-lime, synthetic calcium phosphate or fused silica as fillers [2.59]. Later, quartz because the material of choice due to its inertness and refractive index being compatible with Bowen's resin. It was not until the early 1980s that improved fillers began replacing quartz – they were softer, of smaller particle size and therefore provided composite resins which could be polished. Some of these fillers, including barium and strontium glasses, also provided X-ray opacity which allowed detection of caries, underlying decalcified dentine, marginal overhangs, voids and other defects. There is some concern with regard to the toxicity of the barium glasses, although the quantities employed have not proved a problem to date [2.60]. Hydrolytic instability of barium glasses was found to be a complication in the wet environment of the mouth, resulting in leaching of barium ions and increased degradation of the composite [2.61].

Around this time the first composites appeared which contained colloidal amorphous silica particles [2.62] which are also referred to as pyrolytic or fumed silica particles. Their microscopic size (40–70 nm) gave the resulting composite an essential property which clinicians had demanded, that of polishability. This property coupled with their inertness and low thermal expansion coefficient provided a new generation of anterior filling materials. On the negative side their filler content was considerably lower than the previously produced macrofillers because the particles packed together more closely, creating a much higher surface area per unit volume with a corresponding increase in viscosity of the composite paste. The maximum weight fraction of filler loading was approximately 50% for a workable system. Lack of radiopacity was a further complication.

The mismatch between the thermal expansion of composite resin and the natural tooth led to the use of lithium–aluminium silicate as a filler which in fact shows a coefficient of thermal contraction. Söderholm [2.63] points out that, rather than an advantage, this filler with its negative thermal expansion, when mixed with the positive thermal expansion of the resin matrix, could conceivably produce increased tensile stresses at the filler-matrix interface which may result in debonding.

Although barium provides the best radiopacity for the filler phase of posterior composites, many other high atomic number elements have recently been adopted such as zinc, zirconium, strontium, lanthanum and ytterbium [2.64]. One dental company incorporates ytterbium trifluoride as a radiopaque medium which they claim will release fluoride over a long time span [2.65] thus reducing the likelihood of secondary caries.

2.2.2.1
Manufacture of the Filler

All macrofillers are produced by reducing the size of large particles to within the range 1–30 µm. Hard fillers, like quartz, tended not only to be ground by the process but were also responsible for equipment wear, leading to bead contamination. Another unfortunate consequence was the unavoidable formation of microcracks within the filler particles. The advent of the softer fillers together with improved technology has reduced the average particle diameter to 1–5 µm for modern macrofillers.

Microfine fumed silica fillers are produced by burning silicon tetrachloride in a hydrogen/oxygen atmosphere [2.66]. Alternatively they can be manufactured from colloidal silicon dioxide or by reacting colloidal sodium silicate with hydrochloric acid. The resultant particles are spheroidal in shape and their microsize severely limits their maximum concentration when incorporated into the composite paste, due to the considerable viscosity increase.

It must be remembered that the variation in size between the micro and macro fillers has a profound effect on the properties, and in particular the viscosity, of the resultant composite paste. A macrofiller particle is some thousand times larger than a microfiller particle, but the specific surface area of an equivalent weight of macrofiller is a thousand times smaller, which allows a considerably greater percentage of the macrofiller in the final paste without impairing the handling properties.

The limited filler loading with microfine fumed silica particles was overcome by creating a synthetic filler of organic origin which incorporated the inorganic fumed silica. This has been achieved via two routes.

i) Splintered prepolymer particles; fumed silica particles were mixed with diluted bis-GMA or other less viscous monomers, such as urethane dimethacrylate, and dissolved in a solvent (e.g. chloroform). The stiff immobile quasi-liquid left after pumping off the solvent was heat polymerised and the resultant block was crushed or ball milled to a particle size in the range 20–200 µm (normally a mean particle size value was adjusted to 30 µm). The cured resin can now be regarded as a synthetic filler that possesses none of the uncured resin's disadvantages, but allows a substantial increase in filler loading without a corresponding increase in viscosity. This filler is then mixed with monomer, also containing a limited quantity of fumed silica particles, to produce the commercial paste.

ii) Spherical prepolymer particles; incorporation of pyrolytic silica into partially cured polymer spheres of mean diameter 20–30 µm allowed a greater microfiller concentration, using controlled size distribution, than with the splintered prepolymer type.

A further increase in the fumed silica content has been achieved by controlled agglomeration or condensation of the filler, resulting in homogeneously microfilled composites without a prepolymer system. Another recently developed technique involves sintering the microfiller particles to a porous state followed by grinding to a coarser grade than the original colloidal particles. As much as 70 wt% of filler has been incorporated by this method but it is a moot point as to whether these materials should be classified as microfines.

2.2.2.2
Megafill Inserts

The concept of inserts or "megafillers", which are manufactured in shapes and sizes to almost fill the tooth cavity and thus enhance the composite's mechanical properties, first came to the fore in 1986 [2.67]. These were glass/ceramic materials which, when heat treated, resulted in stabilised beta quartz microcrystals at elevated temperatures [2.68]. They possessed extremely low coefficients of thermal expansion, with radiopacity greater than dentine and even approaching that of enamel. Furthermore introduction of various trace elements such as iron, sulphur and cesium achieved tooth coloured tints which were intrinsically incorporated by controlled heat treatments. Beta-quartz inserts are made from lithium alumino-silicate glass combined with oxide modifiers. The mixture is stirred to produce homogenisation. Molten glass is poured into appropriate predetermined shapes and sizes to fill the bulk of the selected cavities. Surface treatment of the inserts is by organofunctional silane coupling agents, resulting in good bonding between the megafillers and the composite resin. The major advantages of the technique are reduced polymerisation contraction, reduced microleakage, improved dimensional stability, bonding and strength. On packing, the insert is pressed into the cavity which contains uncured composite resin, resulting in good marginal adaptation. With normal composite packing the instruments used tend to pull the resin away from the cavity wall on their removal, initiating the possibility of microleakage sites. This is minimised with the insert technique.

2.2.2.3
Particle Shape

Today most fillers are of an irregular shape with a roughened surface to aid mechanical retention of the matrix phase. The exception are the microfillers which, due to their manufacturing methods are spherical in shape and of smooth surface finish. Spherical particles provide improved packing and reduced stress concentration, when mixed with the resin phase, in comparison with irregular shaped fillers. Reduced mechanical retention of the matrix has been shown by Söderholm [2.69] to be due to their regular shape and smooth surface.

The union between the resin matrix and filler component of any type of composite resin is of paramount importance. Even with the advent of both mechanical bonding and the potentially superior chemical adhesion at the filler-matrix interface, this still remains the weak link in all composite structures. Chemical bonding agents, usually referred to as coupling agents in this context, are therefore an essential component of modern composite technology.

2.2.3
Coupling Agents

Mechanical bonding relies on the surface roughness and imperfections of the filler particles to create undercut areas to retain the resin monomer once it has polymerised. This can be achieved by sintering particles together [2.70] or by etching the glass surface [2.71] to produce a porous surface.

Chemical bonding, as well as being more effective than mechanical interlocking, can, in theory, provide a continuous stress distribution between filler and matrix if the coupling agent possesses properties intermediate between the two phases [2.72]. Hydrolytic degradation at the interface has also been shown to be reduced by the coupling agent [2.73].

The most frequently used coupling agent for dental usage is γ-methacryloxy-propyltrimethoxy silane which is deposited on the filler surface as a 0.025 – 2% aqueous solution sufficient to create at least a monolayer [2.74]. Adhesion is achieved by the siloxane bond between the silanols of the hydrolysed silane and the silanols of the inorganic filler. Adsorption of these silane triols on the filler surface as monomeric or oligomeric layers occurs on completion of the condensation reaction in linkage between the coupling agent molecules themselves and the filler surface via siloxane bonds:

$$
\begin{array}{c}
\text{CH}_3 \\
| \\
\text{H}_2\text{C=C} \qquad\qquad\qquad \text{OCH}_3 \\
| \qquad\qquad\qquad\qquad | \\
\text{COOCH}_2-\text{CH}_2-\text{CH}_2-\text{Si-}\!\mid\!\text{OCH}_3 \;\; + \;\; \text{H}\!-\!\text{O}-\text{Si}\!- \quad \boxed{\text{filler particle}} \\
| \\
\text{OCH}_3
\end{array}
$$

$$\Big\downarrow {}_{-\text{CH}_3\text{OH}}$$

$$
\begin{array}{c}
\text{CH}_3 \\
| \\
\text{H}_2\text{C=C} \qquad\qquad\qquad \text{OCH}_3 \\
| \qquad\qquad\qquad\qquad | \\
\text{COOCH}_2-\text{CH}_2-\text{CH}_2-\text{Si}-\text{O}-\text{Si}\!-\boxed{} \qquad\qquad (2.22)\\
| \\
\text{OCH}_3
\end{array}
$$

The silane film formed possesses a strongly bound polysiloxane network plus hydrolysed silane and small polysiloxane molecules, the latter molecules being susceptible to removal by mouth fluids. The non-homogeneous nature of the film means that, at an interface, the low density material can allow the ingress of fluids [2.75]. To provide acceptable and long term mechanical properties the coupling agent has been shown to require more than a monolayer coverage [2.76].

The bonding between the organic filler of the microfine composites and the matrix phase is incomplete [2.77] because the heat cured organic filler is highly cross-linked, which allows little chance for monomer molecules to diffuse into it and react with any unused double bonds. The limited number which do succeed restrict the number of entanglements and also the micromechanical retention. However, unlike the silane bonded inorganic fillers, the organic bond has some advantageous effects such as bond density increase with reduction in cross-linked density of the organic filler [2.63].

2.2.4
The Curing System

2.2.4.1
Autopolymerising Systems

Since composite filling materials incorporate methacrylate end groups, their method of polymerisation has essentially been that indicated for autopolymerised and heat cured acrylics, as described in Sect. 2.1. Both the original powder/liquid and two paste systems were cured by incorporation of the initiator and activator as separate components which came into contact only on mixing. Although other peroxides like acetyl, alkyl or hydroxy have proved acceptable, benzoyl peroxide remains the initiator of choice. Likewise, the activator commonly preferred by most manufacturers is *N,N*-dihyroxyethyl-*p*-toluidine which possesses superior colour stability to the methyl variety found in denture repair resins [2.78]. These tertiary aromatic amines, with electron donation in the para state, have superior initiation ability to the corresponding tertiary aliphatic amines [2.79] although the latter's colour stability is superior. Brauer [2.79] also considered the condensation resulting from the amine by-products of the redox reaction to be responsible for discoloration. The amine cation can enter into a side reaction with the hydrogen atoms of the aromatic ring which may be the cause of colour changes. Amines substituted in positions 3 and 5 of the phenyl ring have proved much less susceptible to such changes according to Bowen and Argentar [2.80, 2.81]. Examples of 3, 5 substituted amines are *N,N*-diethanol-3,5-di-tertbutyl aniline and *N,N*-dimethyl-3,5-xylidine. Yellowing of the set material by the amine and its by-products is minimised when a UV absorber is part of the composite's make-up.

2.2.4.2
Light Cured Systems

The externally applied light energy can be in the form of ultraviolet (UV) or, more commonly, visible light. The former was used in the early light cured composites when a wavelength around 360 nm provided the maximum initiation energy. Benzoin and its derivatives and benzil ketals were photoinitiators patented by Waller [2.82]. Benzoin methyl ether was a typical example which underwent photofragmentation with the formation of polymerisation-initiating radicals [2.83].

Visible light cured (VLC) resin-composite materials have all but reduced the UV system to obsolescence. They generally employ photosensitised free radical initiators like an α-1,2 diketone such as camphorquinone (CQ) and an amine reducing agent such as dimethylaminoethyl methacrylate (DMAEMA) or dimethyl-*p*-toluide [2.84]. The camphorquinone, with a maximum blue light intensity at 468 nm, has a concentration of 0.17–1.03 mass% of the resin phase, while that of the DMAEMA reducing agent is 0.86–1.39 mass%, according to Taira et al. [2.85]. Combination of the photosensitiser and reducing agent provides an extended absorption range within the visible spectrum.

The photoinitiation mechanism of a camphorquinone/amine system can result in a hydrogen atom being abstracted from the monomer by the excited ketone, or it may be due to an electron transfer reaction from a donor molecule to the excited ketone (excited complex or exciplex), followed by proton transfer [2.86]:

$$
\begin{array}{ccc}
\underset{\substack{R\quad R}}{\overset{\displaystyle O}{\underset{\displaystyle \|}{C}}} & \xrightarrow{h\nu} & \underset{\substack{R\quad R}}{\overset{\displaystyle O^{\cdot}}{\underset{\displaystyle \|}{C}}} \xrightarrow[\text{crossing}]{\text{intersystem}} \underset{\substack{R\quad R}}{\overset{\displaystyle O^{\cdot}}{\underset{\displaystyle \|}{C}}}
\end{array}
$$

ketone	excited singlet	excited triplet

$$
\underset{\text{amine}}{\overset{\displaystyle R\diagdown}{\underset{\displaystyle R\diagup}{\bar{N}-CHR_2}}} \longrightarrow
\left[\; \underset{\substack{R\ \ R\ \ R\ \ R}}{\overset{\displaystyle O\qquad\quad H}{\underset{\displaystyle \|\qquad\quad |}{C\ \leftarrow\ N-CR_2}}}\; \right]^{\cdot}
\xrightarrow[\text{transfer}]{\text{electron}}
\left[\; \underset{\substack{R\ \ R\ \ R\ \ R}}{\overset{\displaystyle O^{-}\qquad\quad H}{\underset{\displaystyle |\qquad\quad\; |}{C^{\cdot}\ +\ {}^{+}N\!-\! CR_2}}}\; \right]
$$

triplet exciplex   hydrogen transfer

$$
\underset{\substack{R\ \ R\quad R\ \ R}}{\overset{\displaystyle OH}{\underset{\displaystyle |}{C^{\cdot}\ +\ \dot{N}-CR_2}}} \tag{2.23}
$$

free radicals

Of the two free radicals formed the amine radical is thought to be a more effective initiator. The excited state wavefunction of the exciplex formed from a donor molecule (A) and an acceptor molecule (Q) can be expressed as:

$$\Psi_E = a\,\Psi_1(A^+Q^-) + b\,\Psi_2(A^-Q^+) + c\,\Psi_3(A^*Q) + d\,\Psi_4(AQ^*) \tag{2.24}$$

The last two terms on the right-hand side represent the excited resonance states whereas the first two terms are the charge resonance states. The electron affinities and relative ionisation potential of the molecules will influence the relative balance between the states in the final wavefunction.

Following photoactivation the reaction stages of the free radicals $(A^{\cdot})$ are as described previously (Scct. 2.1.2), i.e.

Initiation $A^{\cdot} + M \;\rightarrow\; AM^{\cdot}$
Propagation $M_n^{\cdot} + M \;\rightarrow\; M_{n+1}^{\cdot}$
Termination $M_n^{\cdot} + M_m^{\cdot} \rightarrow\; M_{n+m}$

The attenuation of visible light transmission through the composite results in a variable depth of cure, which is one of the limiting factors and the reason why such light cured systems must be packed incrementally into the tooth cavity. If I_0 is the light intensity entering the surface of the composite, I is the light intensity at depth d below the surface and γ is the absorption coefficient of the medium, then Lambert's Law states that:

$$I = I_0 e^{-\gamma d} \tag{2.25}$$

The available light intensity within the medium at depth d is expressed by the transmittance τ, such that

$$\tau = I/I_0 = 10^{-\alpha d} \tag{2.26}$$

where α is the absorbance and $\gamma = 2.303\,\alpha$, 'α' being the linear absorption coefficient. Also, the light intensity absorbed by the photosensitiser across a thickness δd at depth d in the medium is

$$I_n = I\,[1 - e^{-\alpha_s \delta}] \approx I\,\alpha_s\,\delta d \tag{2.27}$$

where α_s is the absorption coefficient of the photosensitiser.
According to Beer's Law

$$\alpha_s\,\delta d = \varepsilon_s\,C_s \tag{2.28}$$

where ε_s is the molar absorptivity of the photosensitiser of concentration C_s at the absorbed frequency.

Thus, in terms of the number of quanta absorbed by the photosensitiser per unit volume at a depth d, the general formula for nonuniform monochromatic light absorption of the composite resin can be expressed as

$$I_\alpha = \varepsilon_s C_s I_0 \cdot 10^{-\alpha d} \tag{2.29}$$

Alternatively, expressed in generalised wavelength (λ) form for interaction of the photosensitiser molecules and the complete composite system we obtain

$$I_\alpha = C_s \int I_0(\lambda)\,\varepsilon_s(\lambda)\,10^{-\alpha(\lambda)d}\,d\lambda \tag{2.30}$$

In general we can assume that C_s is constant when this exists in excess of the photon intensity.

Effective polymerisation requires a minimum light intensity of 500 Wm^{-3} [2.87]. A problem with VLC composites is their sensitivity to the surgery lights if exposure is prolonged, resulting in partial polymerisation which then increases the viscosity of the composites and impairs flow [2.88].

Investigations into the use of coherent monochromatic light (lasers) have shown that, provided their wavelength matches the photosensitiser's absorption spectrum, they can in principle be used as a curing mechanism. However, considerable caution is required at high laser light intensities where non-linear optical effects may affect some organic systems and in particular patient safety must be protected.

2.2.4.3
Dual Cured Systems

To overcome the problem of unreacted monomer within the clinically cured composite resin a technique known as the inlay-onlay extra-orally post-cured system has been devised. The single paste composite incorporates both initiation for light and heat curing mechanisms. In practice a separating medium is applied to the prepared tooth cavity prior to packing the composite. After light polymerisation the inlay is removed and post-cured at an elevated temperature (100–120 °C) for approximately 10 min in a dry heat oven. Since this post-cure approaches the glass transition region of the polymer matrix it was thought by several authors to induce additional polymerisation due to the increased segmental mobility within the resin [2.89, 2.90]. The inlay is finally bonded into the dental cavity with adhesive. Properties such as degree of polymerisation, strength and hardness have been claimed to increase whereas shrinkage and post-operative pain have decreased.

2.2.5
Other Constituents

Stabilisers or inhibitors are commonly employed in minute quantities (300 – 1000 ppm) to alleviate the possibility of prepolymerisation of the resin matrix. As with the unfilled resins the substituted phenols or hydroquinones are most popular.

Pigmentation by oxides of iron for example, is frequently used as it meets the stringent requirements of colour and chemical stability in the hostile oral environment, coupled with a wide range of aesthetic shade control.

Ultraviolet stabilisers are also used to overcome the effects of discolourisation by ultraviolet sources. Weak UV lights, as found in night clubs and discos, cause the composite filling to appear dull and lifeless. Incorporation of optical brighteners provides the composite with a fluorescence to match that of the natural tooth and prevent mismatch in these circumstances.

2.2.6
Classification of Dental Composites

The complexity and variety of composite filling materials available to the profession is now so vast that many authors have attempted to classify them in order to assist the dental practitioner in choosing the correct material for the particular situation. Such classifications are based purely on the material properties of the composite resins. In 1983 Lutz and Phillips [2.91] suggested a classification based on filler type, consisting of three fundamental systems from which all other varieties could be formed. These types were traditional (conventional) macrofillers, microfillers and microfiller complexes as illustrated in Fig. 2.4.

The schematic structure shown for traditional macrofilled composites provides for satisfactory optical and physical properties and, possibly, radiopacity but they lack polishability and have surface roughness leading to plaque accumulation and staining. The difference in hardness of the filler and matrix phases coupled with their large particle size accounts for their poor finishing and roughness characteristics. Modern conventional composites tend to possess smaller (10 to < 3 µm), softer, more rounded particles which pack more efficiently to increase the filler loading (75 wt%) of the paste. The quality of the surface finish, although improved, soon deteriorates due to particle dislodgement [2.92, 2.93].

Hybrids which contain reinforcement of the organic matrix with microfillers have reduced the traditional composites to obsolescence, the differences in

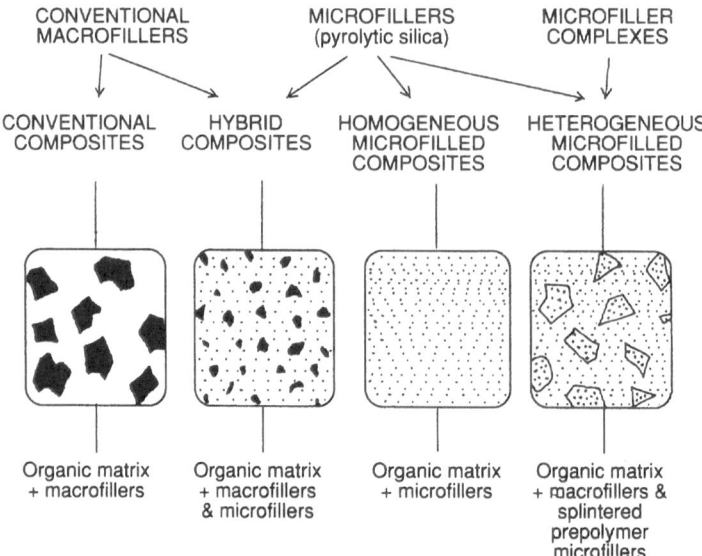

Fig. 2.4. Composite classification adapted from Lutz and Phillips

properties between the matrix and filler phases being reduced for these hybrids. Use of microfillers has led to better viscosity control resulting in enhanced wear resistance, although as anterior restorations they are not ideal presenting a rough surface after time in the hostile oral arena. The reduced macrofiller size (~ 1 μm) combined with the microfiller content, yield superior properties in comparison with the microfiller free conventional systems. Filler content can reach 80–90 wt% (only a small percentage being microfiller) with computer controlled particle size distribution and packing.

The homogeneous microfilled composites corresponding to an organic matrix and directly admixed microfillers exhibit excellent surface smoothness and polishability. Even microfiller removal by attrition in the mouth still leaves a lustrous appearance due to the very small filler particles involved. These small particle size fillers provide the minimal working surface for particle removal which points to improved wear resistance. Unfortunately with such small fillers involved (0.04–0.2 μm) the loading is severely restricted due to the increased viscosity caused by the large surface area which they present. Purely homogeneously microfilled composites are not available commercially but other ways are possible to create increased filler content.

As previously explained in Sect. 2.2.2.1, the novel incorporation of pre-polymerised organic fillers containing inorganic microfillers has considerably raised the filler loading to a maximum of 60 wt%, without adversely increasing viscosity of the resultant paste. These are the so-called heterogeneous micro-filled composites used anteriorly. They have permanent surface smoothness, excellent polishability and aesthetics and good wear resistance providing the concentration of the dispersed microfillers is equal in both the prepolymerised particles and in the polymerisable organic matrix. Under-stress bearing conditions, as in occlusal contact areas, these composites break down at their interfacial junction [2.94]. This inferior interfacial bond is the main cause of their technique sensitivity. Incorrect finishing can lead to fracture of the prepolymerised particles forming fissures along the interfacial surface phase. Additionally, since their filler loading is less than traditional and hybrid systems, polymerisation shrinkage is greater. At present the heterogeneous microfilled complexes containing splintered prepolymerised particles are the most extensively used anterior restoration. Spherical prepolymerised types, although possessing a lower polymerisation contraction due to improved filler loading, are still mainly experimental systems. Similarly, the heterogeneous microfilled resins with agglomerate microfiller complexes have little clinical history in spite of their excellent surface qualities and finishing.

A more recent classification by Willems et al. [2.95] attempts to rank commercially available composites with respect to mean particle size, filler size distribution, filler content, Young's modulus, surface roughness, surface hardness, compressive strength, and scanning electron microscope appearance. From these results the composites were divided into five categories, namely; densified, microfine, miscellaneous, traditional and fibre reinforced as depicted in Fig. 2.5. Composites suitable for posterior restoration should possess Young's moduli equal to or greater than that of dentine according to Nakayama et al. [2.96]. Having assumed a Young's modulus of 18.5 GPa, this was related to an imaginary volume filler percentage of 60% by implementing the phenomenological model of Braem [2.97] for dental composites, where E is the modulus and X is the percentage volumetric filler:

$$E = 3103.33 \exp (0.029771720 \, X) \tag{2.31}$$

Thus the densified class of composites (Fig. 2.5) were divided into compact ($< 60\%$) and midway ($> 60\%$) inorganic filler content. Two further subdivisions, according to mean particle size, resulted in the ultrafine ($< 3 \, \mu m$) and the fine ($> 3 \, \mu m$). Both heterogeneous and miscellaneous composites in the Willems et al. [2.95] classification were subdivided according to the type of prepolymerised or agglomerated fillers used. Analysing their results the authors concluded that the ultrafine compact filled composites showed considerable promise for posterior use due to their high inorganic filler content, greater Young's modulus than dentine, good Vickers hardness value compared to dentine, surface roughness values between 0.48 and 0.71 μm which were similar to enamel, and compressive strengths suggesting posterior occlusal loads would be supported.

Anterior restorations were better served by the ultrafine midway filled type since their small particle size provided low intrinsic roughness and sufficient rigidity to withstand the lower stresses at the front of the mouth. Class IV cavities and large Class III and V restorations are suitable.

The high polishability of the microfines makes them acceptable for small Class III and V restorations but for larger cavities they are contraindicated because of their low modulus and filler content.

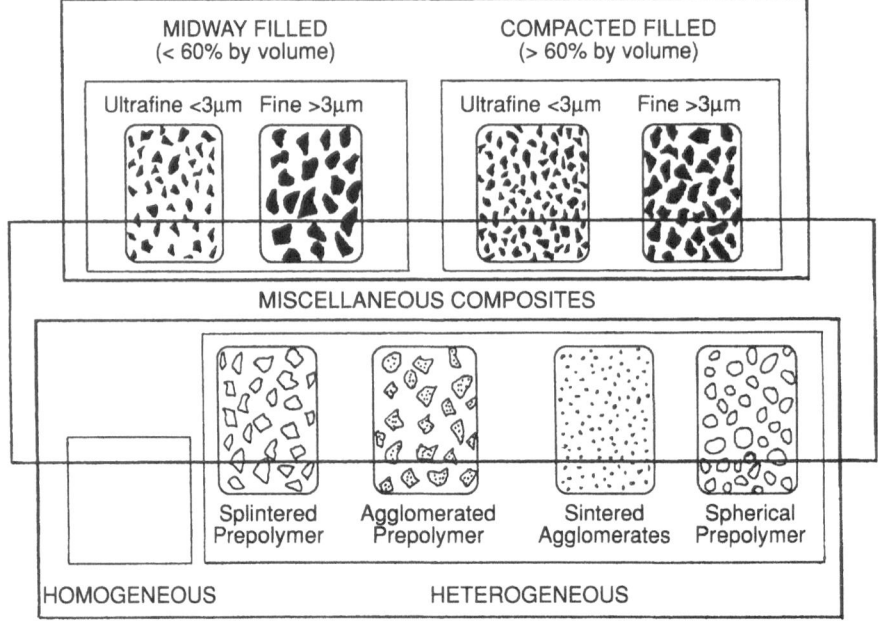

Fig. 2.5. Composite classification adapted from Willems et al.

Larger particle sizes (6–10 µm) for both the fine midway and fine compact types increase wear rates and reduce their suitability for anterior or posterior use, respectively.

Of the miscellaneous composites containing prepolymerised and inorganic fillers the ranking was difficult because of their heterogeneous nature. The large quartz particles in traditional composites result in surface roughness, dullness and lack of radiopacity, rendering these materials obsolete. The fibre reinforced composites possess high Young's modulus but their glass-ceramic fillers are hard, and poor wear properties result, making them less than ideal for posterior situations.

2.2.7
Physical Properties of Composite Resins

Near perfect matching of the subtle shading of natural tooth substance has now been achieved commercially, while at the same time fulfilling the stringent requirements laid down by the various national and international dental authorities for colour and chemical stability in the oral environment. Incorporation of optical brighteners has resulted in fluorescence compatible with natural tooth in daylight and artificial lighting including weak UV lights as encountered in night clubs. However, radiopacity remains a problem with composite filling materials which has only been partially solved.

As with the unfilled resins mentioned in Sect. 2.1.6, the use of heavy metal elements to absorb X-rays does have the undesirable effect of reducing the mechanical properties of the composite. Radiopaque monomers, due to their need for large percentages of bromine or iodine to make them effective, have been unstable and therefore unsuccessful [2.98]. Use of radiopaque glass fillers are much more profitable, although radiopacity is incorporated into the glass in its molten state, which changes the characteristics of the glass in relation to refractive index, hardness and solubility. Reduction in filler/matrix bonding can occur although the silicone content in these radiopaque glasses is limited. Barium sulphate, calcium tungstate and various rare earth elements, some with fluoride incorporation, are the most common radiopaque agents, especially for posterior composites. All anterior prepolymerised composites are radiolucent due to the exclusive use of fumed silica as the microfine filler constituent.

The coefficient of thermal expansion of enamel and dentine is 11 and $8 \times 10^{-6} \, ^\circ C^{-1}$, respectively. Composite resins have considerably higher values which can allow the ingress of fluids and bacteria at the cavity margins when the patient drinks or eats substances at other than mouth temperature. Due to their higher monomer content, the microfilled composites possess somewhat greater thermal expansion coefficients in comparison to their macrofilled and hybrid counterparts. For the same reason, that is the monomer content, microfilled composites have higher thermal conductivity and diffusivity values than other composites. Although composite resins are thermal insulators they are not so efficient as the tooth substance they are replacing. This creates the need for cavity lining materials such as glass ionomers (polyalkenoates) and calcium hydroxides. It should be noted that eugenol-based liners are not recommended due to their staining and inhibitory effects on composite resins in general. Table 2.7 lists some of these physical properties.

Table 2.7. Physical properties of composite resins

Property	Hybrid	Microfine
Thermal Expansion Coefficient ($10^{-6} \, ^\circ C^{-1}$)	30 – 53	50 – 70
Thermal Conductivity ($Wm^{-1}K^{-1}$)	1.0	1.1
Thermal Diffusivity (mm^2s^{-1})	0.68	0.69
Radiopacity (% Al)	200 – 250	–

2.2.8
Mechanical Properties of Composite Resins

In posterior composites under masticatory forces during food consumption, the need for a high Young's modulus to match that of the tooth is particularly essential. Certainly the values for dentine (12–18 GPa) are currently attained by the macrofilled/hybrid systems containing large filler volumes. The much higher modulus of enamel is similar to the fillers themselves, but the monomer phase being some twenty-fold less rigid limits the modulus of the actual composite resin. Thus it is apparent that increase in filler loading enhances modulus. Increased cross link density is also known to raise modulus by reducing the mobility of polymer chains in the network. However, at elevated temperatures the elastic modulus decreases, but high cross link density reduces the demise in the modulus above the glass transition temperature, as can be shown by dynamic thermal analysis.

The need for polishable anterior composites has resulted in reduced filler size in the microfine region, which creates its own problems of decreased filler loading due to the increase in surface area limiting the wetting of the microfine particles by the monomer phase. The Young's modulus of microfilled composites is therefore considerably reduced in comparison with the highly filled hybrids (Table 2.8).

Dynamic mechanical properties can provide some indication of the stability of composite networks. All tests having a compressive component such as elastic modulus, hardness, compressive yield strength, flexural strength and creep resistance correlate well with one another and improve with increased volume filler content. In an ideal system, where maximum adhesion between the filler and the resin matrix is achieved, these compressive properties would predictably increase with the filler concentration in a linear fashion. In practice this is far from true, with modulus, for example, only showing a significant increase when filler loading exceeds approximately 50% of the total mix. Braem et al. [2.99] have confirmed this using acoustic spectroscopy. Furthermore, it appears that the volume of filler present and not particle size or distribution is the main controller of modulus in dental composite systems, together with packing efficiency. Similar filler loading relationships hold for creep and yield strength [2.100]. The ADA specification No. 1 for dental amalgams under a 37 MPa load over a period of 4 h gives very similar results

Table 2.8. Mechanical properies of composite resins

Property	Hybrid	Microfine
Young's Modulus (GPa)	11–20	3–7
Flexural Strength (MPa)	100–170	70–90
Tensile Strength (MPa)	45–70	25–40
Compressive Strength (MPa)	250–400	300–400
Hardness (VHN)	60–120	20–50
Fracture Toughness (MNm$^{-3/2}$)	1.1–1.8	0.8–1.0

for high copper enriched amalgams as found for composite resins in the range 0.06–0.4% creep, with the highly filled hybrids showing the greatest creep resistance. There are suggestions that the marginal fracture associated with posterior composites may in some way be related to the observed creep of these resins.

The intrinsic property, fracture toughness, has become extensively research-ed due to its possible association with wear resistance of dental composites [2.101]. Certainly, fracture toughness does increase with volume filler content, indicating this property is greater for hybrids than for microfilled complexes. The low fracture toughness of microfine complexes may also be partially a consequence of less than ideal interfacial bonding between the prepolymerised organic filler and the matrix, despite the silane treatment of such particles. Wear is such a complex topic that definite correlation between it and other properties like fracture toughness is difficult. We can say with some certainty that observed wear occurs by fracture at the microscopic level followed by macroscopic failure and resultant material loss. Leinfelder [2.102] showed that, in vivo, composites with some large particle size concentration produced generalised wear patterns whereas microfilled resins showed wear loss only at the occlusal contact area. Such localised wear in microfilled composites was related to low fracture toughness and improved fatigue life when tested at low stress levels. At high stress levels, which mimic the clinical situation more accurately, the fatigue life decreased due to embrittlement caused by the faster test rate. Some correlation between wear resistance and fracture tough-ness in vivo by Truong and Tyas [2.103] considers that a static fracture tough-ness test and a dynamic fatigue test together can predict wear resistance. Thus high wear resistance complies with high fracture toughness, small inherent flaw size and high crazing stress. In vivo observations appear to be best re-presented by a three-body wear test using various abrasive agents under high loads. Composites with high fracture toughness and modulus but low average particle size produce the best correlation between in vitro testing and in vivo observations.

Unlike the above-mentioned mechanical properties, tensile strength is dependent on the matrix phase rather than the filler loading. Macrofilled/hybrid resins possess higher tensile strengths when tested, which is contrary to theoretical predictions that point towards higher tensile strength with smaller particle size. Poor filler/matrix bonding may account for this discrepancy. Table 2.8 presents some of these mechanical properties.

2.2.9
Chemical and Biological Properties of Composite Resins

One major concern with all dental composites is their polymerisation shrinkage on curing. High filler volume content minimises the problem since the actual monomer present on conversion to polymer is the cause. Although contraction is undesirable, it is to some extent nullified by the oral environment since, given time, the composite resin can absorb solvents, resulting in a degree of expan-sion. Microfilled systems absorb approximately 2.5 times more water than the macrofilled type due to their greater resin volume. A large proportion of this liquid absorption occupies voids left by free monomer and oligomers on their elution, a consequence of which is that expansion is restricted and does not completely rectify the total volume contraction. Use of hydrophilic monomers [2.104] or SOC, as mentioned previously, can provide some solace. Post cured onlay/inlay systems, having their final cure at elevated temperatures outside the mouth, reduce the quantity of solvent uptake and hence lessen the chance of solvent absorption as the means of combating shrinkage. Results indicate that the chemistry of the monomer is significantly more effective in controlling water uptake than the degree of conversion. For example, monomers containing ether linkages or hydroxyl groups absorb larger quantities of water than those with no such hydrophilic moieties [2.105].

Creep behaviour, being a time-dependent phenomenon, has been investigated with composites stored in ethanol over a six-month period [2.44]. Since this alcohol is a good solvent for dental composites, a catastrophic increase in creep resulted, indicating the time-dependent demise in related properties such as modulus, compressive strength, flexural strength and fracture toughness. Over the same time-scale water immersion showed only small changes. Ethanol was chosen because its solubility parameters are similar to that of bis-GMA and acids which are present in plaque, suggesting creep and related properties would decline on extended periods in the mouth. Microfilled systems showed the greatest decline in such mechanical properties on account of their high monomer content and large interfacial area at the resin-filler junction.

Polymerisation shrinkage may be one of the most important factors in determination of the longevity of dental composites due to the internal stresses created. Either adhesive and/or cohesive failure takes place. Adhesive failure results in the resin matrix shrinking away from the cavity wall and away from the filler particles. Cohesive failure is associated with voids or microcracks within the resin itself. Such a highly stressed situation may be the precursor to microleakage and corrosion. Composites have been shown to exhibit volume contractions in the range 1.67–5.68% [2.106] with microfilled resins using chemical curing systems providing the highest shrinkages. It is well known that chemically activated resins shrink towards the centre of the material on setting, whereas light cured systems contract towards the surface which is exposed to the light source. Both cases result in internal stresses within the restoration, which cause microcracks and imperfect marginal seals. In the situation where the filling material is packed against the tooth wall, forces exerted at the interface are considerable when the composite shrinks on curing. These forces have been measured by various authors. Values of 2.8–3.9 MPa were recorded by Davidson and De Gee [2.107] for conventional and microfilled composites, while Bowen et al. [2.108] reported larger values of 6.1–6.4 MPa for a microfilled resin and 5.5–7.3 MPa for a conventional composite. Contraction gaps supposedly appear at the tooth/composite interface when no bonding agent is present, but in the situation where an effective bonding resin is used, Davidson and De Gee [2.107] predicted the composite would flow during polymerisation and cohesive bonding would be retained. It has been observed clinically that large composite fillings, particularly in premolars, cause the cusp enamel to crack on a horizontal plane parallel but some 2–3 mm below the occlusal table. This has been confirmed to be due to the contraction stresses created when the composite polymerises. If not treated, this can lead to post-operative pain and microleakage.

Microleakage around the cavity margins remains a possible site for ingress of liquids and bacteria. However, for bacteria cell penetration the gap must be approximately 2 µm. These gaps have been measured and are particularly problematical at the junction between the dentine of the cervical margin and the composite resin. The bond to enamel being stronger than that to dentine results in the resin pulling away from the weaker dentine bond as the restorative material shrinks on curing. This creates a space at the cervical margin of large proportions with respect to bacteria cell dimensions. Even worse, the modern light-activated dental composites contract towards the light source and therefore away from the cervical margin on setting. Implications for development of recurrent caries, discoloration, pulpal irritation and thermal sensitivity are thus increased. The use of the acid etch technique for enamel coupled with a bonding agent for dentine minimises these undesirable possibilities, especially if incremental packing is used. Table 2.9 lists some of these foregoing properties.

The water absorption characteristics of composite resins are well documented (Braden and Clarke [2.109], Kalachandra and Wilson [2.110], for example). From the biological viewpoint, water transport properties can contribute to the release of free monomer with consequent toxic effects, leading to possible rupture of the resin/matrix interfacial bonds [2.111]. However, water uptake can relieve areas of stress concentration and act beneficially in counteracting the setting shrinkage of composites on polymerisation [2.112]. Since the resin monomer phase is responsible for water diffusion processes, as verified by Braden and Clarke [2.109], the microfines with a lower filler content have higher diffusion coefficients and equilibrium water uptake than both hybrid and conventional composites. Furthermore, water uptake has considerably deleterious effect on the mechanical properties of composite resins, in particular the microfine variety. Both yield stress and fracture toughness have been shown to decrease by up to 30%, which clinical observations of fracture in restorations show might be linked to composite embrittlement due to water uptake [2.113].

Table 2.9. Chemical/biological properies of composite resins

Property	Hybrid	Microfine
Volume Shrinkage (%)	2.5–3.7	3.0–5.0
Diffusion Coefficient in water (37 °C)		
Sorption (10^{-6} mm^2 s^{-1})	1.9–3.1	3.4–8.1
Desorption (10^{-6} mm^2 s^{-1})	2.8–9.1	10–30
Equilibrium Uptake (%)	0.8–2.3	3.5–5.1

2.3
References

2.1.   Crawford JWC(1932) US Patent 2042458: British Patent 405699 (ICI)
2.2.   Bolker HI (1974) Natural and synthetic polymers: an introduction. Dekker, New York
2.3.   Margerison D, East GC (1967) An introduction to polymer chemistry. Pergamon, Oxford
2.4.   Allinger NL, Cava MP, DeJongh DC, Johnson CR, Lebel NA, Stevens CL (1976) Organic chemistry, 2nd edn. Worth, New York
2.5.   Flory PJ (1953) Principles of polymer chemistry. Cornell University Press, Ithaca, New York
2.6.   McCabe J F, Wilson SJ, Wilson HJ (1978) Br Dent J 144:167
2.7.   De Clerck JP (1987) J Prosthet Dent 57:650
2.8.   Harvey W (1966) Br Dent J 121:49
2.9.   Sainsbury PR (1964) Dent Pract Dent Res 14:243
2.10. Davy KWM, Causton BE (1982) J Dent 10:254
2.11. Davy KWM (1994) European Patent 0684222 A 1
2.12. Lambrecht JR, Kydd WL (1982) J Prosthet Dent 12:865
2.13. Ruyter IE, Espevik S (1980) Acta Odontol Scand 38:169
2.14. Sauder W, Paffenbower GC (1942) Physical properties of dental materials. US Government Printing Office, Washington DC, pp 166–167
2.15. Crank J, Park, JS. (1968) Diffusion in polymers. Academic Press,London, New York, pp 259–289
2.16. Crank J (1975) The mathematics of diffusion, 2nd edn. Claredon Press, Oxford, p 48
2.17. Braden M (1975) Polymeric prosthetic materials. In: von Fraunhofer JA (ed) Scientific aspects of dental materials. Butterworths, London, Boston, Chap. 15
2.18. Ostromislensky I (1927) US Patent 1 613 673
2.19. Turley SG (1968) Appl Poly Symp 7:237
2.20. Rodford RA (1988) The development and evaluation of high impact strength denture base materials (thesis). University of London, London
2.21. Buchdahl R, Nielsen LE (1950) J Appl Phys 21:482
2.22. Merz EH, Claver GC, Baer M (1956) J Poly Sci 22:325
2.23. Schmitt JA, Keskkula H (1960) J Appl Poly Sci 3:132
2.24. Bucknall CB, Smith RR (1965) Polymer 6:437
2.25. Kambour RB (1962) Nature 195:1299
2.26. Berry JP (1961) J Poly Sci 50:107
2.27. Newman S, Strella S (1965) J Appl Poly Sci 9:2297
2.28. Bucknall CB (1977) Toughened plastics. Applied Science Publishers, London
2.29. Bowman AJ, Cook M, Jennings EH, Rannie I (1974) J Dent Res 53:1080
2.30. Schreiber CK (1971) Br Dent J 130:29
2.31. Skirvin DR, Vermilyea SG, Brady RE (1982) Military Med 147:1037
2.32. De Boer J, Vermilyea SG, Brady RE (1984) J Prosthet Dent 51:119
2.33. Yazdanie N, Mahood M (1985) J Prosthet Dent 54:543
2.34. Capaccio G, Ward IM (1973) Nature Phy Sci 243:143
2.35. Ladizesky NH, Ward IM (1983) J Mater Sci 18:533
2.36. Braden M, Davy KWM, Parker S, Ladizesky NH, Ward IM (1988) Br Dent J 164:109
2.37. Ladizesky NH, Ward IM (1995) J Mater Sci Mater Med 6:497
2.38. Gutteridge DL (1988) Br Dent J 164:177
2.39. Braden M, Clarke RL, Pearson GJ, Cambell Keys W (1976) Br Dent J 141:269
2.40. Patel MP, Braden M (1991) Biomaterials 12:645
2.41. Pearson GJ, Picton DCA, Braden M, Longman C (1986) Inter Endodont J 19:121
2.42. Bowen RL (1962) US Patent 3 066 112
2.43. Bowen RL (1965) US Patent 3 179 783
2.44. Ferracane JL (1989) Trans Acad Dent Mater 2:6

2.45. Cowperthwaite GF, Foy JJ, Malloy MA (1981). In: Gebelein CG, Koblitz FF (eds) Polymer science and technology, vol 14. Plenum Press, New York
2.46. Ferracane JL (1983) The correlation between the physical properties and degree of conversion in unfilled bis-GMA-based resins (PhD dissertation). Northwestern University, Chicago, Illinois
2.47. Braden M, Davy KWM (1986) Biomaterials 7:474
2.48. Espe Fabrik Pharmazeutischer Praparate, Gmbh (1972) British Patent 1 263 541
2.49. Foster J, Walker RJ (1974) US Patent 3 825 518
2.50. Ruyter-IE, Sjøvik IJ (1981) Acta Odontol Scand 39:133
2.51. Schrollenberger CS, Dinbergs K (1961) SPE Trans 1:31
2.52. Arroyo Dental Products (1978) US Patent 4 119 610
2.53. Antonucci JM, Griffith JR, Peckoo RJ, Termini DJ (1979) J Dent Res 58: Abstract 599
2.54. Thompson VP, Williams EF, Bailey WJ (1979) J Dent Res 58:1522
2.55. Stansbury JW (1991) J Dent Res 70:527 Abstract 2088
2.56. Millich F, Jeang L, Byerley TJ, Eick JD (1991) J Dent Res 70:527 Abstract 2086
2.57. Byerley TJ, Eick JD, Chen GP, Chappelow CC, Millich F (1992) Dent Mater 8:345
2.58. Eick JD, Robinson SJ, Byerley TJ, Chappelow CC (1993) Quintessence Int 24:632
2.59. Bowen, RL (1979) J Dent Res 58:1493
2.60. Stanley HR, Bowen RL, Folio J (1979) J Dent Res 58:1507
2.61. Söderholm KJM (1983) J Dent Res 62:126
2.62. Jørgensen KD, Hørsted P, Janum O, Krogh J, Schultz J (1979) Scand J Dent Res 87:140
2.63. Söderholm KJM (1985). In: Vanherle G, Smith DC (eds) Posterior composite resin dental restorative materials. Peter Szulc, Netherlands
2.64. Willems G, Noack MJ, Inokoshi S, Lambrechts P, Van Meerbeek B, Braem M, Roulet JF, Vanherbe G (1991) J Dent 19:362
2.65. Ott G (1990) Ivoclar-Vivadent Report, Liechtenstein Nos
2.66. Degussa (1979) Grundlagen und Anwendungen von Aerosil Schriltereihe Pigmente, Nr. 11, Degussa aerosil fumed silica. Degussa, W. Germany
2.67. Bowen RL, Setz LE (1986) J Dent Res 65:797 Abstract 642
2.68. Bowen RL (1988) US Patent 4 744 759
2.69. Söderholm KJ (1979) Effects of water on particle reinforced composites for dental use (M.Phil. thesis). Univ of Sussex
2.70. Ehrnford L (1976) Odontol Revy 27:51
2.71. Bowen RL, Reed LE (1976) J Dent Res 55:748
2.72. Kwei TK (1965) J Polym Sci 3:3229
2.73. Söderholm KJ (1981) J Dent Res 60:1867
2.74. Plueddemann EP (1982) Silane coupling agents. Plenum, New York
2.75. Bascom WD (1972) Macromolecules 5:792
2.76. Johannson OK, Stark FO, Vogel GE, Fleschmann RM (1967) J Comp Mater 1:278
2.77. Lambrechts P, Vanherle G (1983) General Dentistry 21:276
2.78. Asmussen E (1980) Acta Odont Scand 38:95
2.79. Brauer GM (1981). In: Gebelein CG, Kablitz FF (eds) Polymer science and technology, vol 14. Plenum, New York
2.80. Bowen RL, Argentar H (1967) J Amer Mater Assc 75:918
2.81. Bowen RL, Argentar H (1971) J Dent Res 50:923
2.82. Waller DE (1973) US Patent 3 709 866
2.83. McGinnis VD (1975) J Radiat Curing 2:3
2.84. Dart EC, Nemcek J (1975) British Patent 1 408 265
2.85. Taira M, Urabe H, Hirose T, Wakasa K, Yomaki M (1988) J Dent Res 67:24
2.86. Watts DC (1992) Setting Mechanisms of Dental Materials-Symposium. Loch Lomond, Scotland
2.87. Watts DC, Amer O, Combe EC (1984) Br Dent J 156:209
2.88. Dionysopoulos P, Watts DC (1990) J Oral Rehabil 17:9
2.89. Wendt SL (1987) Quintessence Int 18:265
2.90. Burke FJT, Watts DC, Wilson NHF, Wilson MA (1991) Br Dent J 170:269

2.91. Lutz F, Phillips RW (1983) J Prosthet Dent 50:480
2.92. Lambrechts P, Vanherle G (1982) J Oral Rehabil 9:169
2.93. Heuer GA, Garman TA, Sherrer JD, Williams HA (1982) J Amer Dent Assc 105:246
2.94. Schaefer HWR (1981) Quintessence 32:1685
2.95. Willems G, Lambrechts P, Braem M, Celis JP, Vanherle G (1992) Dent Mater 8:310
2.96. Nakayama WT, Hall DR, Grenoble DE, Katz JL (1974) J Dent Res 53:1121
2.97. Braem M (1985) An in-vitro investigation into the physical durability of dental composites (PhD thesis). Katholieke University Leuven, Belgium
2.98   Tsunekawa M, Ishibashi M (1984) British Patent 2 181 144 A
2.99   Braem M, Lambrechts P, Vanherle G (1988) Trans Acad Dent Mater Proc, Gaitherburg, MD. USA
2.100 Ferracane JL (1985) Dent Mater 1:11
2.101 Lloyd CH, Mitchell L (1987) J Oral Rehabil 11:257
2.102 Leinfelder KF (1987) Int Dent J 37:152
2.103 Truong VT, Tyas MJ (1988) Dent Mater 4:318
2.104 Bowen RL, Rayson JE, Dickson G (1982) J Dent Res 61:654
2.105 Kawaguchi M (1988) J Japan Soc Dent Mater Rev 7:143
2.106 Goldman M (1983) Aust Dent J 28:156
2.107 Davidson CL, De Gee AJ (1984) J Dent Res 63:18
2.108 Bowen RL, Nemoto K, Rapson JE (1983) J Amer Dent Assoc 106:475
2.109 Braden M, Clarke RL (1984) Bionaterials 5:369
2.110 Kalachandra S, Wilson TW (1992) Biomaterials 13:105
2.111 Braden M (1977) Dental Update 1:369
2.112 Bowen RL, Rapson JE, Dickinson J (1982) J Dent Res 61:654
2.113 Ferracane JL, Antonio RC, Matsumoto H (1987) J Dent Res 66:1140

CHAPTER 3

Elastomeric Materials

Elastomeric or rubber-like materials are materials or polymers whose inter- and intra-molecular forces are sufficiently low for the free energy of deformation to be mainly a function of entropy. In the first part of this chapter, hydrocolloids are discussed, which have rubbery compliance because they contain water; in the second part, plasticised glassy polymers are considered these also having a rubber-like compliance.

Originally, rubber technology was based on natural rubber, but from the time of the first world war onwards, a highly diverse synthetic rubber industry has developed. However the principle polymers used in the rubber industry for tyres, engineering and domestic use are not those used in dentistry. This is because the materials used in the dental surgery and dental laboratory need to be processable by extremely simple technology, e.g. hand mixing, and sometimes curing at room temperature. This, and the requirements concerning toxicity, restricts the materials that can be used.

3.1
Impression Materials

M. Braden

Many procedures in restorative dentistry require an impression to be made of a dentate or edentulous mouth, in order for an appliance or component (denture, partial denture, crown, bridge, inlay or orthodontic appliance) to be fabricated. Obviously it is essential that the appliance fits well, and in many applications this means the permitted dimensional changes in the impression material are limited to 0–0.15% [3.1.1]. Since the preparation of an appliance or component often involves the stages of impression taking, casting up a Gypsum model, investment and casting, it is clear that there are considerable demands on each stage.

The first stage is the taking of an impression, and hence dimensional stability is a prime requirement, as well as the more obvious requirement of strength. The other feature is that the impression is taken with a material that is initially fluid, but sets (usually by a chemical reaction) in situ; hence the set material has to be removed over undercuts. As well as strength, elastic recovery (permanent set) is also of major importance; the other aspect of dimensional stability being that the material should not change dimensions on storing: it may be a week between impression taking and subsequent casting of a model.

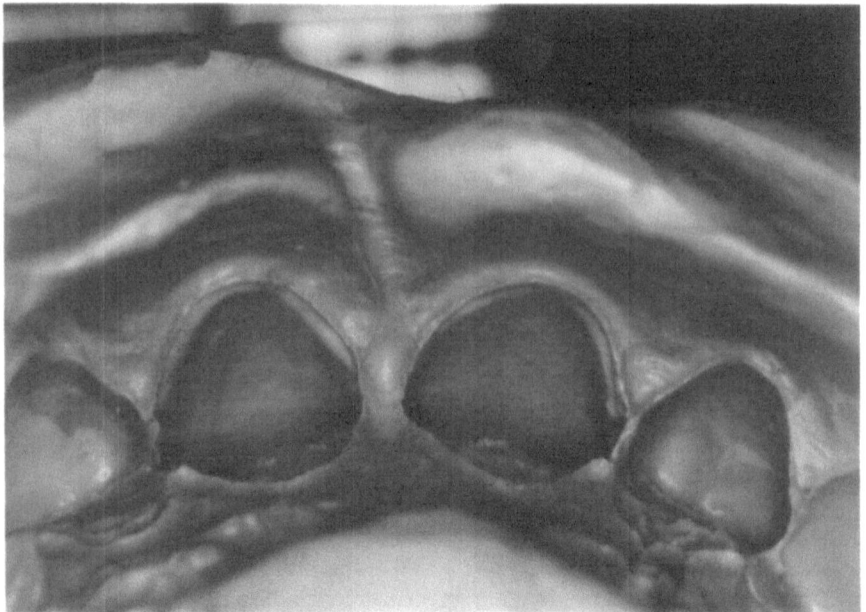

Fig. 3.1.1. An impression using a silicone impression rubber by a two-stage technique

The current clinical scene demands that strict precautions are taken to avoid cross-infection. Therefore an impression material must withstand immersion in a suitable sterilising solution without adverse effects on dimensional stability or any other property.

Finally, the material must not produce any adverse effects on the patient, even though the material is only in the mouth for 5–10 min.

The materials described in this section are commonly called "Elastic Impression Materials" by the dentist. This is something of a misnomer, and compliant materials with rubber-like moduli are actually involved. They are all inserted into the mouth as a fluid material, which changes to a viscoelastic solid in 3–10 min. This change is usually effected chemically, but in one case by a thermally reversible gel.

Figure 3.1.1 shows an impression after the material has been removed; it is held in a so-called impression tray. The example shown has involved a two-stage impression technique; a low viscosity material has first been syringed round the teeth (the purple material), and this has been backed up by a very viscous material in the impression tray (the brown material). It will be seen how the fluid material has reproduced the detailed structure of the teeth.

3.1.1
Hydrocolloids

3.1.1.1
Reversible Hydrocolloids

These materials are based on Agar-Agar, a polysaccharide extracted from certain types of seaweed [3.1.2].

Chemically it is a sulphuric ester of a linear polymer of galactose. It is a thermally reversible gel, which is a fluid at 50 °C, but a solid at mouth temperature. Its handling is somewhat cumbersome, necessitating a water bath to keep the material, usually dispensed in small syringes, in a fluid condition. Impression trays are of a cored metal construction, through which water is circulated when the impression is seated, to accelerate gelling. The impression must be removed with care, because it has low tear resistance. Also, for work of any precision, a model must be cast immediately, because the loss of water by syneresis is rapid, resulting in rapid shrinkage and distortion. The resulting impression has excellent surface detail. However, these materials have been largely superseded by synthetic elastomers

3.1.1.2
Irreversible Hydrocolloids-Alginates

The alginates used in dentistry are largely sodium or potassium alginate, although triethanolamine alginates are occasionally used. A sodium alginate solution is rapidly converted to a gel by the addition of a divalent ion such as calcium, which cross-links the system (Fig. 3.1.2) [3.1.3].

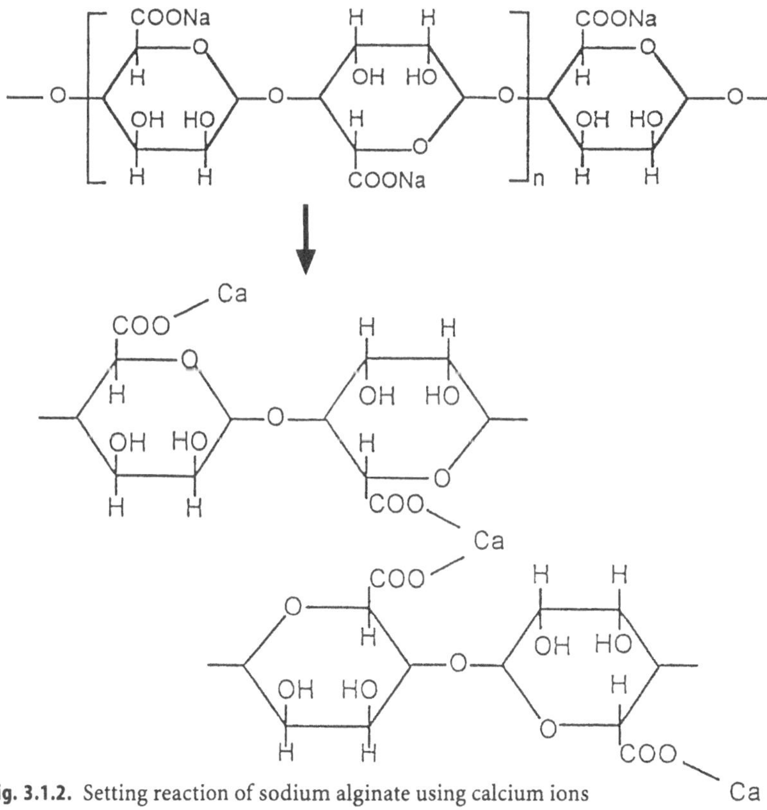

Fig. 3.1.2. Setting reaction of sodium alginate using calcium ions

An alginate impression material is dispensed as a powder, containing sodium alginate, calcium sulphate as a sparingly soluble source of calcium ions, a retarder such as sodium trisulphate, a reinforcing filler such as diatomaceous earth, other ingredients to control pH, and colourants. The dentist adds water to this in a measured amount, which is mixed to form a paste, and this is inserted in the mouth on an impression tray.

The function of the sodium phosphate is to remove calcium ions to delay cross-linking and give time for mixing and placement of the impression:

$$CaSO_4 \longrightarrow Ca^{2+} + SO_4^{2-}$$

$$Na_3PO_4 \longrightarrow 3Na^+ + PO_4^{3-}$$

The calcium and phosphate ions combine and precipitate insoluble calcium phosphate; this process continues until the sodium phosphate is exhausted, and then calcium ions commence the cross-linking reaction. Clearly other divalent salts could be used; indeed, in the past lead silicate was used!

When first mixed, the alginate paste is alkaline, but changes during setting towards pH 7, or sometimes less. The final pH is of consequence, because it can affect the surface of the gypsum model against which it is cast.

Alginate impression materials are relatively cheap, easy to use, and have controlled setting behaviour. They are used for first impressions and applications where high accuracy is not required, and are extensively used for such applications. However, like reversible hydrocolloids, they have poor dimensional stability and tear strength. If immersed in aqueous solutions, they swell initially, then contract as soluble salts are extracted and the chemical potential reversed. Hence immersion in sterilising solutions must be of limited duration. A useful review of sterilising impression materials as a whole has been made [3.1.4].

3.1.2
Elastomers

The advent of fluid, room temperature vulcanising (RTV) elastomers since the 1950s has greatly extended the range of elastic impression materials; subsequent advances in such materials have greatly benefited clinical dentistry. They are usually dispensed as two-paste systems, one paste containing the fluid polymer plus fillers and additives, and the other the cross-linking system. Various viscosity grades are available, this being controlled by the filler (and, in the case of polysulphides, plasticisers). The dentist mixes these just before use, setting behaviour being adjusted to give the dentist ample working time.

Mostly these materials are variants on industrial polymers, but one material (an imine terminated polyether) was developed specifically for dentistry [3.1.5].

3.1.2.1
Condensation Silicones

The RTV polymers used in this type of impression material are α,β-hydroxy terminated polydimethylsiloxanes first described by Nitzsche and Wick [3.1.6].

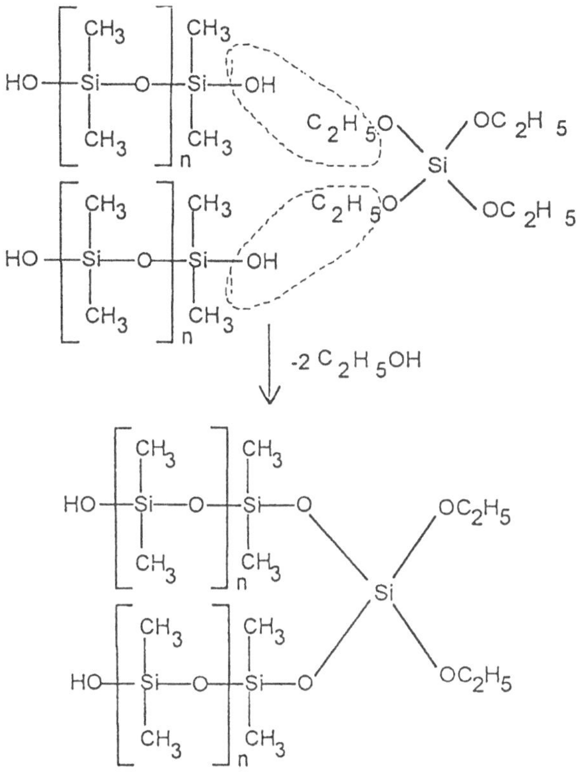

Fig. 3.1.3. Setting reaction of a condensation silicone impression polymer

They are cross-linked via chain ends by an alkoxysilicate, accelerated by an organo-tin compound, such as stannous octoate, with the evolution of an alcohol (Fig. 3.1.3). Subsequent evolution of the alcohol by evaporation contributes to the shrinkage of such materials, which can amount to 0.1–0.4%, depending on the filler content. Braden has shown that this process can be quantified by application of diffusion theory [3.1.7]. One potential problem with these materials is the limited shelf life of the cross-linking system.

They are greatly superior to hydrocolloids in dimensional stability and strength, and have also been used as soft lining materials (see Chap. 3.2).

3.1.2.2
Addition Silicones

In this case the polydimethylsiloxane is terminated with vinyl groups [3.1.8], and these are cross-linked by silicones containing –SiH groups, with a platinum salt as a catalyst (Fig. 3.1.4). In this case, no by-products of the reaction are formed, and these materials have the best dimensional stability of all elastic impression materials. They do however have disadvantages, as follows.

i) It is now standard clinical practice for dentists to wear rubber gloves, consequent on the proliferation of the AIDS virus. Often the more viscous

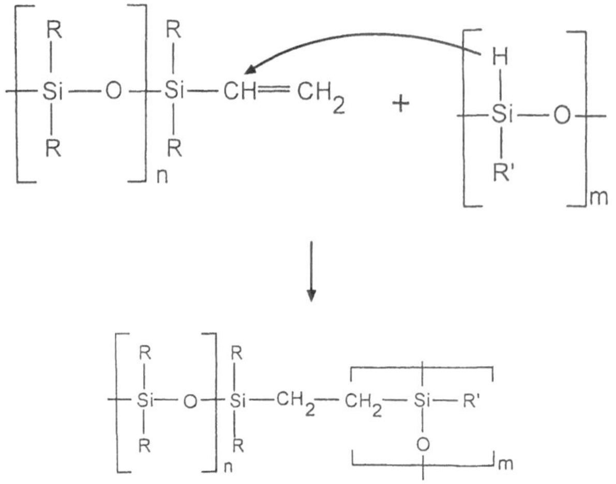

Fig. 3.1.4. Setting reaction of an addition silicone impression polymer

grades (the so-called Putty types) are mixed by kneading in the hand. If the gloves are the natural rubber latex type, depending on the vulcanisation system, the sulphur in the rubber poisons the platinum salt catalyst, and setting is inhibited.

ii) If the Gypsum model is cast up immediately after the impression is taken, then there is a reaction between the water and residual Si–H groups, with hydrogen evolution resulting in a porous model. Hence casting should be delayed for at least 20 min.

iii) Addition silicones tend to be weaker mechanically than corresponding condensation materials, although they are clinically adequate.

iv) Both addition and condensation silicones are of course hydrophobic. This can lead to defects in the impression in contact with blood or saliva. Some types now contain a detergent to improve wettability.

3.1.2.3
Polysulphides

Dental impression materials of this type are usually based on the Thiokol LP-2 type of material, fluid polysulphides first described by Jorczak and Fettes [3.1.9]. The general structure is;

$$HS-[CH_2-CH_2-O-CH_2-CH_2-S-S-]_n-CH_2-CH_2-O-CH_2-CH_2-SH$$

This polymer contains 2 mol% of branched, thiol terminated groups to facilitate cross-linking. The setting process consists of chain lengthening and cross-linking, usually by an oxidising agent, which in dentistry is mostly lead dioxide, PbO_2 (Fig. 3.1.5). The rate of the reaction is very sensitive to the presence of moisture. In spite of the apparently simple chemistry, a given and reputable proprietary impression material can at times be quite capricious in its setting behaviour.

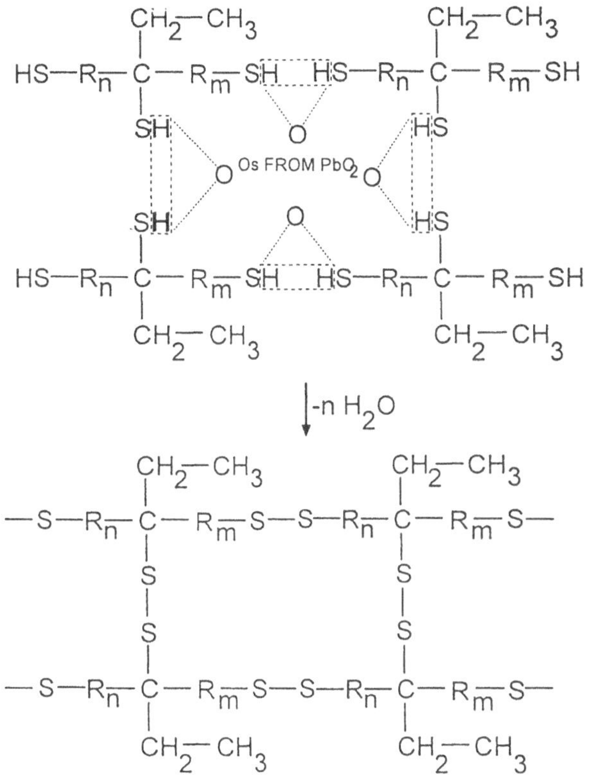

Fig. 3.1.5. Setting reaction of a polysulphide impression polymer

Other cross-linking agents have been tried in dentistry, notably hydroperoxides, but without success [3.1.10]. One proprietary product used a copper salt cross-linking reagent [3.1.11], the mechanism of which is obscure. Polysulphides can also be cross-linked with disulphides by an exchange with terminal and pendant SH groups of the polymer:

$$2 \; \sim\!\!\sim\!\!\sim \; SH + R\text{--}S\text{--}S\text{--}R \;\rightarrow\; \sim\!\!\sim\!\!\sim \; S\text{--}S \; \sim\!\!\sim\!\!\sim \; + 2\,RSH$$

This is a reversible reaction, which is quenched by the inclusion of, for example, zinc oxide which reacts with the RSH compound as it is formed.

The advantages of polysulphide impression materials are:

i) acceptable dimensional stability, which was only bettered with the advent of the Addition Silicone materials;
ii) the highest strength of all impression materials.

The disadvantages are:

i)   they are rather slow setting;
ii) their permanent set is inferior to all types of silicone polymers, except those which are heavily loaded with filler;
iii) not surprisingly, they are rather odiferous materials, that are rather messy to handle.

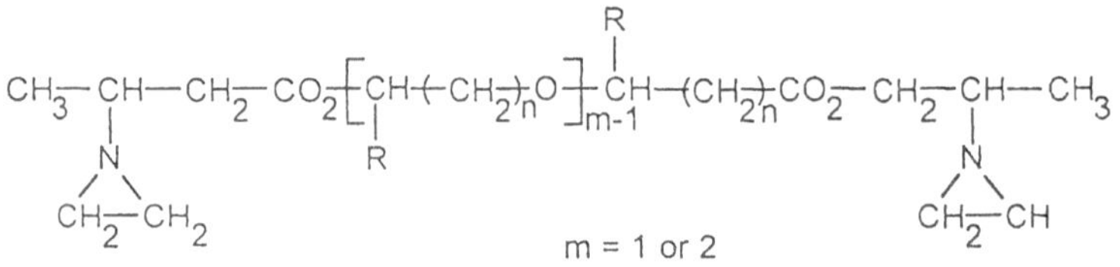

$$CH_3-CH-CH_2-CO_2\left[CH-(CH_2)_nO\right]_{m-1}\overset{\overset{R}{|}}{CH}-(CH_2)_nCO_2-CH_2-CH-CH_3$$

(with substituents: left side N bonded to CH_2-CH_2; center R; right side N bonded to CH_2-CH)

m = 1 or 2

Fig. 3.1.6. Generalised structure of an imine terminated polyether impression polymer

Nevertheless, until the advent of polyether and addition silicone materials, they were the material of choice, certainly in the UK and the USA.

3.1.2.4
Imine Terminated Polyethers

These materials are extremely interesting in that they were specifically developed for dentistry by Espe Pharmazeutischer Fabrik GmbH in Germany [3.1.5].
The general chemical structure is given in Fig. 3.1.6; (the parent polymer is
synthesised from a hydroxy-terminated poly ether, which is esterified with an
unsaturated acid, and ethylene imine is added to the double bond). The setting
reaction is via a cationic ring opening mechanism involving the imine ring.
Hence there are no by-products of the cross-linking reaction.

Also, cross-linking is across chain ends, resulting in a rather hard set material
which can be difficult to remove from large complex impressions.

Although with the other materials discussed previously there are many versions available from many companies, there is basically only one manufacturer
of the polyether material, and one consistency grade available.

This material is easy to use, quick setting, with adequate strength, although
the high hardness referred to above can cause sufficient stress on removal to
make rupture a possibility. However the use of proper technique by the dentist
should avoid this.

The dimensional stability of the material is adequate, but its hydrophilicity
can cause distortion if left in contact with water for any length of time, although
this effect is less pronounced with later versions. Clearly this is of importance in
the matter of sterilisation [3.1.4]. Some hydrophilicity is advantageous in that it
facilitates the wetting of the oral tissues; see the problems already referred to
with silicone polymers.

Another problem with earlier versions of the material was the possibility of
hypersensitivity of the dentist, his assistant, or the patient to the sulphonic acid
ester used originally as a source of R^+ [3.1.13]. Again, later versions have used an
alternative (unidentified) initiator.

3.1.2.5
A Photo-Polymerising Impression Material

All of the elastomeric materials so far discussed are, to a greater or lesser extent,
in common use, certainly in the western world. One interesting material has

been developed by the DeTrey Dentsply Company in the USA. There is little published on its chemistry, but is said to be a "polyether urethane dimethacrylate". One can speculate on its possible chemistry. Starting with a hydroxy-terminated polyether (compare to the iminepolyether):

$$HO-(R-O)_n-OH$$

The terminal groups could presumably be coupled to hydroxyethylmethacrylate with a di-isocyanate, to give urethane dimethacrylate end groups by inclusion of a suitable initiator system, such as those used in dental composite filling materials (see Sect 2.4.2.2). The dentist can then use the light sources used for the latter materials.

Cross-linking is across chain ends, reminiscent of the imine-terminated polyether.

Clearly, the material must be translucent, making the choice of fillers critical, and special transparent impression trays must be used. This material is not available in the UK at the time of writing.

3.1.2.6
The Future?

With a comprehensive range of materials already available, it is interesting to speculate on whether other materials can (or should!) be developed. Some years ago the author did develop, in collaboration with the chemical industry, a proto-type material based on maleic anhydride terminated low molecular weight poly-butadiene, which was cross-linked in an esterification reaction with diols or triols. This material appeared to show considerable promise. Whilst this has not been commercialised in dentistry, it does indicate that there are always possibilities for the future.

3.1.3
Physical Properties

3.1.3.1
Rheology

As already pointed out, impression elastomers are usually dispensed as two pastes, which have to be mixed in the surgery by the dentist or his assistant. Depending on the clinical circumstances, the required consistency may be any-thing from a putty-like material to a material that can be syringed into the inter-stices of the oral structures.

Rheological studies [3.1.14] have shown these pastes, not surprisingly, to be shear-thinning, and possibly thixotropic materials. Viscosities are in the range of 100–1500 Pa s. Viscosity is largely governed by filler content.

3.1.3.2
Stiffness of the Set Material

In terms of hardness, values between 28 and 80 IRH are encountered, again governed by filler content, i.e. there is generally a correlation between viscosity and hardness. In modulus terms the shear moduli [3.1.15] vary between 0.2 and 3 MPa; corresponding Young's moduli will therefore be ~ 0.6 – 9 MPa. Some of the high moduli are for the so-called "putty" type, a "back-up material" which does not enter the fine detail of the dentition. However, the imine polyether referred to earlier (Sect 3.1.2.4) has an IRH value of ~ 70, which can lead to difficulties in large, highly undercut impressions.

3.1.3.3
Viscoelastic Properties

Ideally, an impression material should recover from the deformations involved on impression withdrawal. Inevitably, with viscoelastic materials, recovery is time-dependant, and there may be some permanent deformation. International specifications indeed stipulate tension and compression set requirements; these are empirical rather than fundamental. Measurements [3.1.15] of $\tan \delta$ have shown values from 0.01 to 0.4. Interestingly, the same work showed a linear correlation between $\tan \delta$ and compression set, but none with tension set. This was attributed to the fact that the viscoelastic measurements were carried out in shear (torsional pendulum), and the set test involves substantial shear deformation; indeed, removal of an impression also involves substantial deformation [3.1.15].

3.1.3.4
Strength Properties

It is axiomatic that an impression should not tear on removal. (Indeed, weaker materials such as alginates do occasionally tear on removal.) In extension to break terms, values of > 50 % are required. All elastomeric impression materials meet this.

Typical tearing energies (in Jm^{-2}) are:

- alginates 200 – 600;
- silicones/polyether 400 – 1200;
- polysulphides 800 – 1700.

These were evaluated [3.1.16] by the "trouser test-piece" first described by Rivlin and Thomas [3.1.17].

3.1.4
Biological Properties

The sterilisation of impressions to avoid cross-infection has already been referred to. It is axiomatic that the impression material should not adversely affect the patient's oral mucosa, albeit that the time of contact is only a matter of minutes. Likewise it should not affect the dentist or ancillary staff, who are in continual contact with the material. Sydiskis has studied the cytotoxicity of impression materials [3.1.18].

3.1.5
General Conclusions

There are now a wide range of impression materials available to the dentist, and while they are not completely ideal, they have nevertheless been developed to a high degree of sophistication.

3.1.6
References

3.1.1. Dental Elastic Impression Materials. ISO Specification 4823.
3.1.2. Phillips RW (1990) Skinners Science of Dental Materials, 9th edn. WB Saunders, Philadelphia, p 111
3.1.3. Braden M (1975) Impression Materials. In: von Fraunhoffer JA (ed) Scientific Aspects of Dental Materials. Butterworths, London
3.1.4. Owen CP, Goolam R (1993) Int J Prosthodont 6:480
3.1.5. Purrman R, Schmidt W (1964) British Patent 1 044 753. Espe Fabrik Praparate
3.1.6. Nitzsche S, Wick M (1957) Kunststoffe 47:433
3.1.7. Braden M (1992) Biomaterials 13:333
3.1.8. Watt JAC (1970) Chem Brit 6:519
3.1.9. Jorczak JS, Fettes EM (1950) Indust Eng Chem 43:324
3.1.10. Braden M (1977) Dental Update 1:489
3.1.11. Molnar E (1964) British Patent 951 472. Coe Laboratories
3.1.12. Braden M, Causton BE, Clarke RL (1972) J Dent Res 51:889
3.1.13. Nally FN, Storrs J (1973) Brit Dent J 134:244
3.1.14. Braden M (1967) J Dent Res 46:429
3.1.15. Braden M, Inglis AT (1986) Biomaterials. 7:45
3.1.16 Braden M (1963) Dent Pract 14:67
3.1.17. Rivlin RS, Thomas AG (1953) J Polym Sci 10:29
3.1.18. Sydiskis RJ, Gerhardt DE (1993) J Prosthet Dent 69:431

3.2
Soft Prosthesis Materials

S. Parker

Commonly, this type of material is used to provide a soft lining for dentures (Fig. 3.2.1). They are used for patients that cannot tolerate a hard denture for various reasons e.g. when the tissues of the denture bearing area show evidence of atrophy. They can also be used to engage undercut areas to aid denture retention without causing trauma to the soft tissues. Other uses include intra- and extra- oral prostheses e.g. obdurators (Figs. 3.2.2 and 3.2.3). They have potential use as soft ear mould materials (Fig. 3.2.4) and mouth guards.

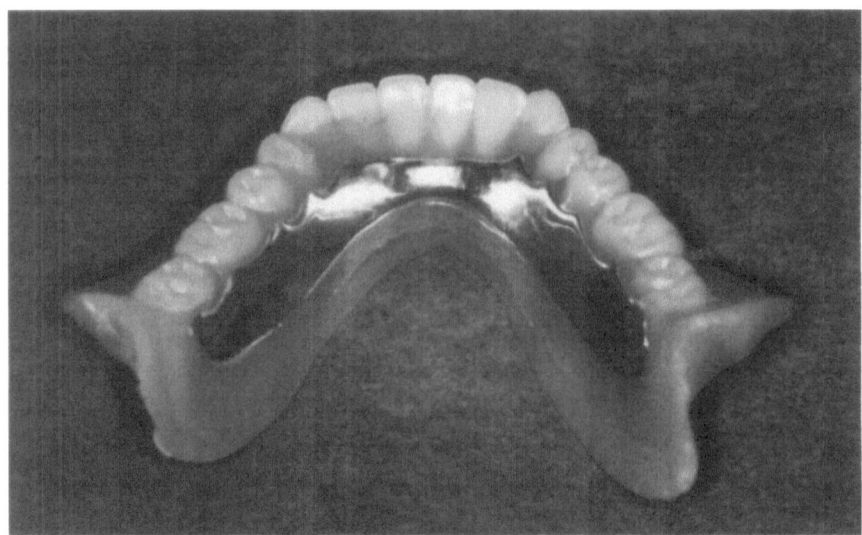

Fig. 3.2.1. A full lower denture with a soft lining

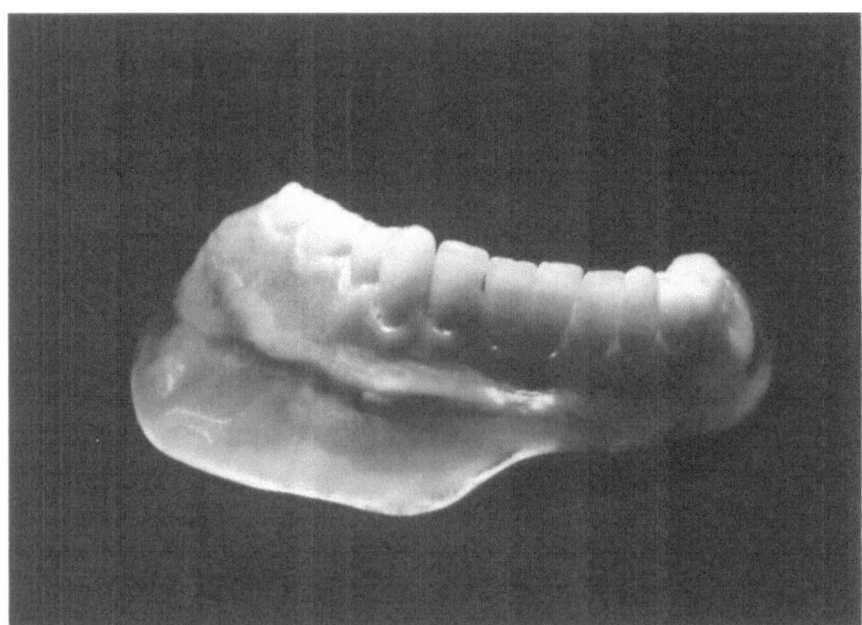

Fig. 3.2.2. A full lower denture plus sulcus deepening

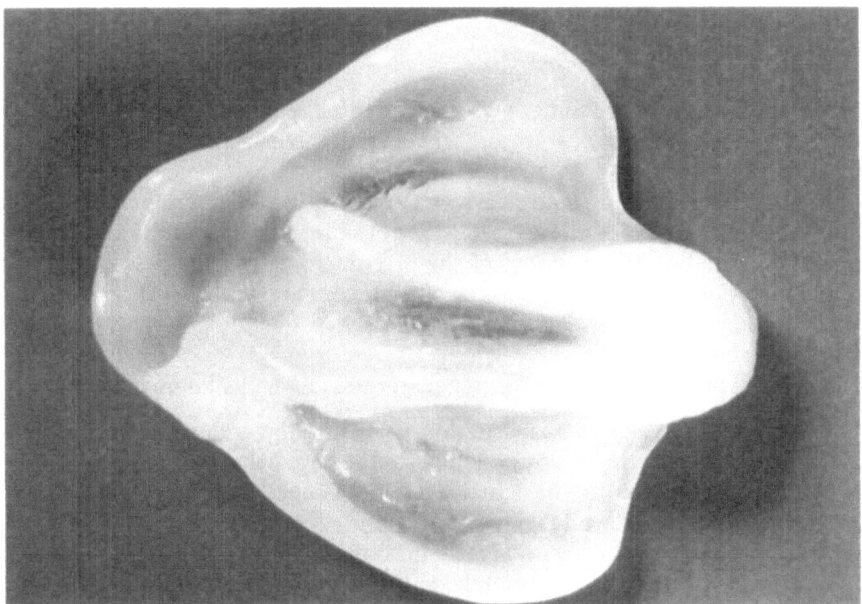

Fig. 3.2.3. A cleft palate appliance

Fig. 3.2.4. Direct insert ear moulds

3.2.1
Soft Lining Materials

Materials of this type are sometimes referred to as "resilient" lining materials. As this term in no way infers "softness" and current materials exhibit a wide range of resilience values it is somewhat erroneous.

In function, soft lining materials serve to distribute the forces of mastication more evenly and absorb energy. They do not reduce the transmitted force, but result in a smaller displacement of the oral mucosa. The stiffness of the soft lining material is less than that of the oral mucosa and will absorb more of the energy and deform more. The energy is released as the lining returns to its pre-deformed shape.

The ideal properties of a soft lining material include the following.

1. Biocompatibility – non-toxic and non-irritant to the oral tissues.
2. Permanently compliant – soft enough for the comfort of the patient.
3. Permanently resilient – the actual level of resilience required is unknown as shown by the range available.
4. Adhere to the denture base and remain so in the mouth.
5. Low water uptake – a high uptake will cause distortion and may result in fouling of the lining due to ingress of bacteria etc.; a similar level to that of the denture base is ideal ~ 2 – 3 %.
6. Wetted by saliva – a thin film of saliva is necessary for the retention of the denture and to act as a lubricant to prevent irritation of the mucosa.
7. Should not support the growth of *Candida albicans* – this is a common yeast responsible for causing denture stomatitis, a common occurrence in denture wearers.
8. Sufficient mechanical properties – enough to withstand normal handling, brushing etc.
9. Easy to clean – not adversely affected by denture cleansers, not stained easily etc.

The majority of soft lining materials currently available can be classified into two main types, soft acrylic and silicone rubber.

3.2.1.1
Soft Acrylic Materials

3.2.1.1.1
Heat Polymerised

These are methacrylate-based systems where the glass transition temperature (T_g) has been reduced by the addition of an external plasticiser. The plasticiser acts as a lubricant between the polymer chains, enabling them to move over each other and so deform more easily. The secondary inter-molecular forces between the polymer chains also have the power to hold other molecules as well as acting as cohesive forces. When a solvent, or other low molecular weight substance, is added to a polymer it is attracted to the chains by these forces and gradually pushes the chains apart. If a large amount is added, the polymer will ultimately

Tabel 3.2.1. Composition of a soft acrylic soft lining materal

Powder	Liquid
Poly(ethyl methacrylate) or related copolymer e. g. with butyl or methyl methacrylate	Higher methacrylate monomer (e. g. butyl, ethyl or 2-ethoxyethyl methacrylate) Plasticiser, usually a phthalate (e. g. dibutyl phthalate or butyl phthalyl butyl glycollate) Cross-linking agent (e. g. ethyleneglycol dimethacrylate)

pass into solution, but if only a small amount of a non-volatile solvent is used, plasticisation occurs.

They are presented as a methacrylate polymer powder with a liquid comprising a methacrylate monomer plus a plasticiser. When the two are mixed they form a dough in the same way as the poly(methyl methacrylate) denture base materials (Chap. 2) and can then be moulded and heat cured. The initiator is the residual benzoyl peroxide present in the polymer powder. Table 3.2.1 gives examples of composition of such materials. The chemical formulae of two of the most commonly used plasticisers are shown below:

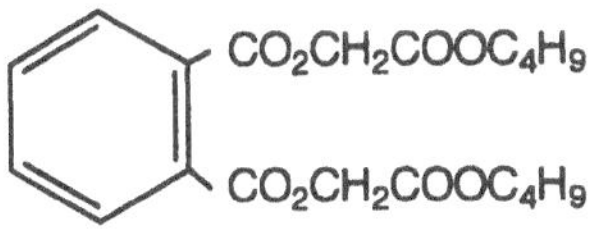

Butyl phthalyl butyl glycollate

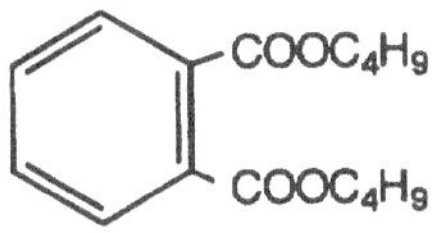

Dibutyl phthalate

Early materials were based on poly(methyl methacrylate) which has a T_g of 125 °C. A large amount of plasticiser was required to produce a material with a T_g lower than mouth temperature. Higher methacrylates are now used for both polymer powder and monomer so less plasticiser is necessary. Table 3.2.2 gives the T_g values of some higher methacrylates. The choice of polymer powder is limited to those that will form a free-running powder at room temperature, i. e. have a T_g above room temperature. Of the monomers listed, 2-ethoxyethyl methacrylate has now been withdrawn due to possible toxicity [3.2.3]. There is now also concern about the possible toxic effects of the phthalate esters used as plasticiscrs [3.2.4]. However, butyl phthalyl butyl glycollate is preferable as it is less easily hydrolysed.

This type of material is well known for its gradual hardening in the mouth due to loss of plasticiser [3.2.5] but it does adhere well to the denture base. Litchfield and Wood [3.2.6] patented a material that reportedly overcame the problem of hardening by using di-2-ethylhexyl maleate as a plasticiser:

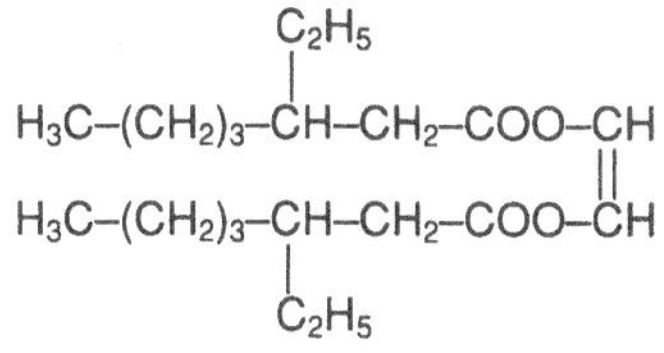

Di-2-ethylhexyl maleate

Table 3.2.2. Glass transition temperatures (T_g) of some poly(alkyl methacrylate)s

Monomer		T_g of polymer (°C)	
Ethyl methacrylate	atactic	65 (range 47–70)	[3.2.1]
	isotactic	12	[3.2.1]
	syndiotactic	66	[3.2.1]
n-Propyl methacrylate		38 (range 35–72)	[3.2.1]
Butyl methacrylate	atactic	20	[3.2.1]
	isotactic	−24	[3.2.1]
Iso-butyl methacrylate randóm		53	[3.2.1]
	isotactic	8	[3.2.1]
	80% syndiotactic 20% isotactic	53	[3.2.1]
Hexyl methacrylate		− 5	[3.2.1]
2-Ethylhexyl methacrylate		−10	[3.2.1]
Decyl methacrylate		−70	[3.2.1]
2-Ethoxyethyl methacrylate		−15	[3.2.2]

As the molecule is unsaturated it is likely to undergo some polymerisation. The double bond has a much lower reactivity than the methacrylate monomer so it is unlikely to copolymerise. However, even a small degree of polymerisation will result in an increase in size which will reduce the tendency for the plasticiser to leach out.

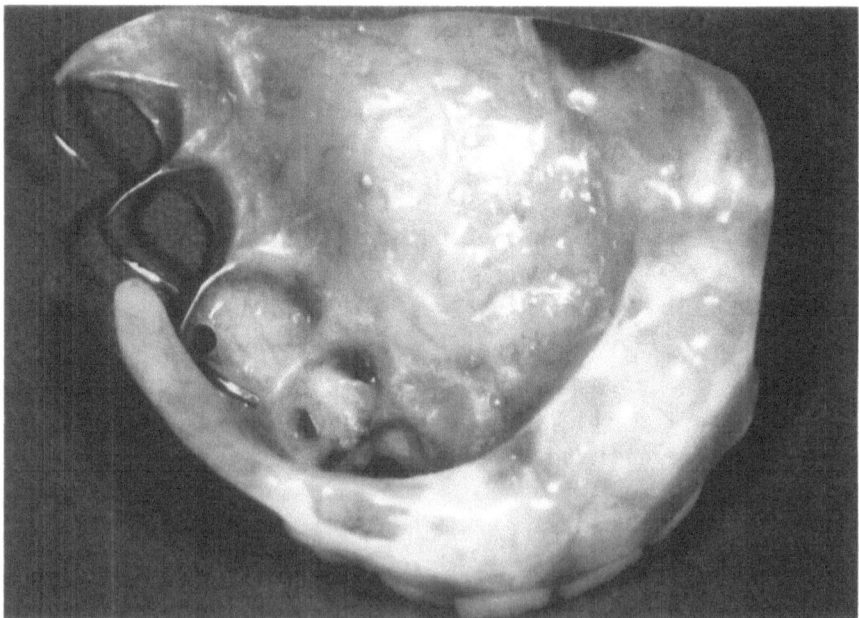

Fig. 3.2.5. An experimental soft lining after 6 months clinical trial

Although seemingly successful, this material was withdrawn as the composition included 2-ethoxyethyl methacrylate. Subsequent work using the same plasticiser has proved it is virtually inextractable [3.2.7]. This system comprised an 80/20 butyl/ethyl methacrylate copolymer powder with 1-tridecyl methacrylate monomer and the maleate plasticiser. Unfortunately the material failed in clinical trials due to excessive water uptake (see Fig. 3.2.5). This will be discussed in a later section.

3.2.1.1.2
Room Temperature Polymerising ˙

Room temperature polymerising versions can be produced (as with the methyl methacrylate based materials described in Chap. 2) by the inclusion of an aromatic amine (e.g. *N, N*-dimethyl-*p*-toluidene shown below) in the monomer liquid:

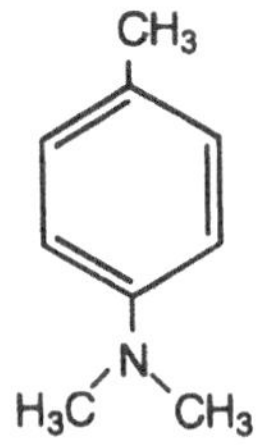

N, N-dimethyl-*p*-toluidine

Generally, heat cured versions are preferred as the polymerisation process is more efficient; room temperature curing can result in a higher level of free monomer remaining in the materials. The presence of the free monomer can result in inferior mechanical properties and reduced biocompatibility. The only advantage of these over the heat polymerised type is convenience.

This type of material is often referred to as "temporary" rather than "permanent" as are the heat polymerising type. However, it is important not to confuse them with tissue conditioner materials which are also sometimes referred to as temporary lining materials and will be dealt with in a later section.

3.2.1.2
Silicone Rubber Materials

3.2.1.2.1
Room Temperature Polymerising

This type is basically the same as the condensation silicone impression materials (see Sect. 3.1), i.e. a two-paste system based on a hydroxyl-terminated poly (dimethyl siloxane) with an alkoxy orthosilicate cross-linking agent and organo-tin catalyst.

The main problem with this type of material is lack of bonding to the PMMA denture base and as such they are usually supplied with an adhesive. This comprises a silicone polymer in a solvent, but even with its use the degree of bonding is still less than that of the soft acrylic materials [3.2.8]. They also have a number of other drawbacks, such as supporting the growth of *Candida albicans* [3.2.9] and having a large water uptake [3.2.5]. One such material, Flexibase (Flexico Developments Ltd., England) was found to have an uptake of ~ 50 wt%! This is somewhat surprising as silicone rubbers are hydrophobic. However, transport of water through them is rapid and the high uptake is due to a combination of this and the presence of hydrophilic filler particles commonly used in silicone rubber materials [3.2.10].

3.2.1.2.2
Heat Polymerising

One such material is Molloplast B (Karl Huber GmbH & Co. KG, Germany); it is a one-paste system of complex chemistry and is derived from a patent by Van Handel [3.2.11]. Table 3.2.3 shows the composition as given in the patent. It appears to be quite straightforward. However, one of the components (RTV108) is a commercial product in its own right [3.2.12]. It is a moisture cure silicone sealant material comprising a number of different components including a hydroxyl terminated polydimethyl siloxane, fumed silica filler, methyl triacetoxysilane and dibutyl tin dilaurate (or other organo-tin catalyst). It will also contain a "process aid" containing multi functional organosilicones. Obviously the paste has to be mixed under dry conditions to prevent the RTV 108 polymerising. During the mixing there must be a reaction to produce a stable paste. It is suggested that the acetoxy-Si groups from the RTV 108 react with the methoxy –Si groups from the silane A 174. However, the situation is a little more complex as the A 174 is only added as a minor component (~ 1.22%) and it also has a role in cross-linking and bonding to the denture base.

Molloplast B, unlike the room temperature polymerising silicones, does adhere to the denture by way of the methacrylate group on the silane (A 174). However, as cross-linking of the gel occurs on storage, adhesion can be variable. This has been improved by use of an adhesive, probably just the silane A 174. Molloplast B also has low water uptake, however it does have low tear energy which is a common problem with silicone rubbers (see later).

Flexor (Weil Dental GmbH, Germany) is another heat cured silicone rubber material presented as a one paste system. It is said to be a polydimethysiloxane-perfluoralkanol dimethacrylate. The paste has a similar odour to that of Molloplast B, so it is thought to be of similar basic chemistry. It too has low water uptake.

Table 3.2.3. Composition of Molloplast B [3.2.11]

Material	Quantity
General Electric RTV 108 (Acetoxy-cure silicone material)	720 parts
Poly(methyl methacrylate)	19.15 parts
γ-Methacryloxypropyltrimethoxysilane	9.16 parts
Titanium dioxide + washung red	0.075 parts

3.2.1.2.3
Acetoxy-Type

This type of material is of the moisture-cure type where the cure is accelerated by the application of heat. Its composition is similar to that of RTV 108 already described in the previous section. The simplified curing chemistry is shown below:

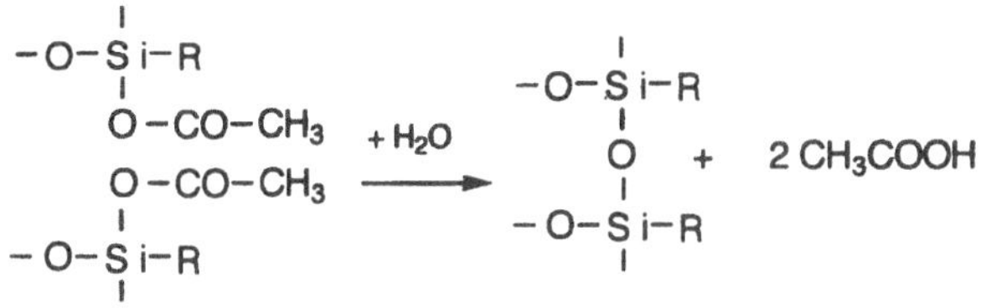

$$
\begin{array}{c}
-O-\underset{|}{\overset{|}{Si}}-R \\
O-CO-CH_3 \\
O-CO-CH_3 \\
-O-\underset{|}{\overset{|}{Si}}-R
\end{array}
\; + H_2O \;\longrightarrow\;
\begin{array}{c}
-O-\underset{|}{\overset{|}{Si}}-R \\
O \\
-O-\underset{|}{\overset{|}{Si}}-R
\end{array}
\; + \; 2\,CH_3COOH
$$

R = Alkyl group

When compared with a number of other commercial materials in a series of in vitro tests, one such material was deemed to have the best combination of properties [3.2.13]. Unfortunately that has since been withdrawn.

An RTV Dow Corning Medical Grade Adhesive type A was investigated as a potential soft lining materials and found to have favourable properties [3.2.14]. More recently, RTV 106 (GEC Silicones Division, USA), which is of similar composition to that of RTV 108, has been investigated as a potential soft lining material and found to compare well with leading commercial soft lining materials [3.2.15].

3.2.1.2.4
Addition Silicone

Results have been presented [3.2.16] for a two-paste silicone rubber material based on an addition reaction rather than a condensation reaction. One paste contains dimethyl vinylsiloxy poly dimethyl siloxane and chloroplatinic acid as the catalyst and the other contains polyhydromethyl siloxane. The pastes are mixed, moulded and then heat cured at 100 °C. The reaction is shown below:

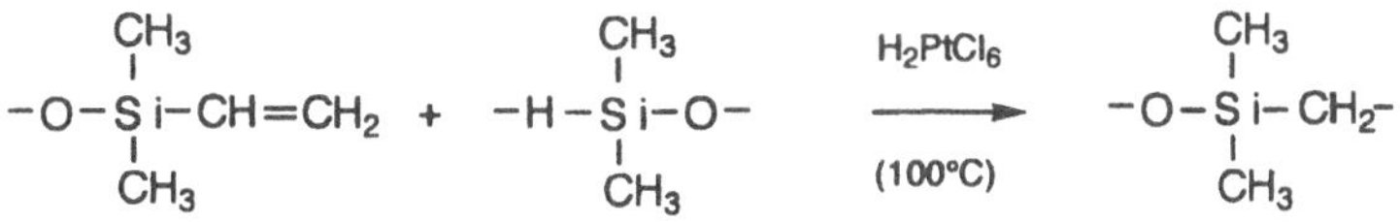

$$
\underset{\overset{|}{CH_3}}{-O-\overset{CH_3}{\overset{|}{Si}}-CH=CH_2}
\; + \;
\underset{\overset{|}{CH_3}}{-H-\overset{CH_3}{\overset{|}{Si}}-O-}
\quad\xrightarrow[\;(100°C)\;]{H_2PtCl_6}\quad
\underset{\overset{|}{CH_3}}{-O-\overset{CH_3}{\overset{|}{Si}}-CH_2-}
$$

R = Alkyl group

Bonding to the PMMA denture base is achieved using a mixture of the silicone base paste and catalyst with denture base dough (i.e. PMMA beads doughed with MMA). This is placed between the silicone and denture base and forms an interpenetrating network (IPN) layer which bonds to both silicone and denture base.

3.2.1.3
Elastomer/Methacrylate Systems

The combination of an elastomer with a methacrylate monomer should produce a material with good strength and adhesion to the PMMA denture. In essence it is a soft acrylic material with no external plasticiser. One such material is commercially available Novus (Hygenic Corp); a one-paste system containing a poly(fluoralkoxy) phosphazine elastomer(PNF 200, Firestone).

$$\left[-N = P \begin{array}{c} OCH_2CF_3 \\ | \\ | \\ OCH_2(CF_2)_x\text{-}CF_2H \end{array} \right]_n$$

PNF 200

The material is based on a patent [3.2.17] and the general composition is given in Table 3.2.4. The paste is produced by milling the elastomer with the methacrylate monomers, initiator etc. As cross-linking will occur, albeit very slowly, even at low temperatures the paste needs refrigerating or even freezing! The paste is rather stiff and does require careful moulding and its elastic nature requires the uncured material to be left in the mould to relax prior to heat curing.

Table 3.2.4. Composition of Novus [3.2.17]

Material	Quantity
Polyphosphazine fluoroelastomer (PNF-200 from Firestone)	100 parts
Trimethylolpropane trimethacrylate (Cross-linker)	18 part
Ethylene glycol dimethacrylate (Cross-linker)	2 part
Poly(methyl methacrylate) (Filler)	10 part
BaSO₄ (Radiopaque filler)	15 part
Lauroyl peroxide (Initiator)	1 part
CdSSe dark red (Pigment)	0.2 part

More recently [3.2.18] softer materials have been produced by blending the phosphazene polymer with a methacryloxypropyl terminated polydimethyl siloxane:

or a substituted phosphazene trimer:

Both act as plasticisers to the system.

Elastomers can be produced in powdered form by cryogrinding at liquid nitrogen temperatures or by latex spraying. It is then necessary to add a partitioning agent to prevent agglomeration. A number of elastomers have proved to be compatible with methacrylate monomers [3.2.19]. Results have been presented for such powdered elastomer/methacrylate monomer systems. One such system comprised a butadiene/acrylonitrile copolymer with 2-ethoxyethyl methacrylate monomer and proved to have excellent physical properties. Unfortunately, as previously stated, 2-ethoxyethyl has now been prohibited for intra-oral use [3.2.3] and free acrylonitrile was also a concern.

Another system, based on a butadiene/styrene block copolymer with *n*-hexyl or ethyl hexyl methacrylate monomers was developed [3.2.20]. This was a one paste system which could be heat cured at 100 °C. Again, physical properties were favourable with the exception of high water uptake. Poly(*cis*-isoprene) is another elastomer which has potential use in this area [3.2.21].

3.2.1.4
Fluoroethylene Copolymers

A heat cured material, Kurepeet from Kureha Chemical Ind. Co., Tokyo comprises a copolymer of 55% vinylidene fluoride, 25% chlorotrifluoro ethylene and 20% tetra fluoroethylene. The resulting material shows good bonding and low water uptake. Clinical trials over two years have shown it to be a successful material [3.2.22].

$$\left[\begin{array}{c}\overset{\displaystyle H}{\underset{\displaystyle H}{\overset{|}{\underset{|}{C}}}}-\overset{\displaystyle F}{\underset{\displaystyle F}{\overset{|}{\underset{|}{C}}}}\end{array}\right]_n \qquad \left[\begin{array}{c}\overset{\displaystyle Cl}{\underset{\displaystyle F}{\overset{|}{\underset{|}{C}}}}-\overset{\displaystyle F}{\underset{\displaystyle F}{\overset{|}{\underset{|}{C}}}}\end{array}\right]_n \qquad \left[\begin{array}{c}\overset{\displaystyle F}{\underset{\displaystyle F}{\overset{|}{\underset{|}{C}}}}-\overset{\displaystyle F}{\underset{\displaystyle F}{\overset{|}{\underset{|}{C}}}}\end{array}\right]_n$$

Poly(vinylidene fluoride)   Poly(chlorotrifluoroethylene)   Poly(tetrafluoroethylene)

Visible light cured versions have been developed based on the combination of fluoroalkyl methacrylate monomers with a vinylidene fluoride/hexafluoropropylene copolymer or a vinylidene fluoride/hexafluoroethylene/hexafluoropropylene copolymer. The materials produced were wettable, had low water uptake and low residual monomer [3.2.23].

3.2.1.5
Natural Rubber

The earliest soft lining materials included natural rubber (*cis*-1,4-polyisoprene) cross-linked with sulphur and had a very short intra-oral life. High water uptake caused distortion and they became foul after 1 – 2 months in the mouth. Adhesion to the denture base also became a problem when PMMA replaced vulcanite (a hard natural rubber) as the preferred denture base material. A system using a natural rubber/PMMA graft copolymer with a zinc dimethyl dithiocarbamate-sulphur curing system that cures at 100 °C was developed [3.2.13]. It still required use of an adhesive (the graft copolymer in solution) to bond to the denture base although it had excellent mechanical properties. Unfortunately bonding was weak and water uptake still high. It was abandoned because of the potential adverse reaction to the dithiocarbamate [3.2.24].

A synthetic polyisoprene is being investigated in the elastomer/methacrylate monomer formulations as described earlier [3.2.21]

cis-1,4-Polyisoprene

3.2.1.6
Vinyl Resins

Poly(vinyl chloride) was one of the first synthetic resins to be used as a soft lining material [3.2.25]

$$\left[\begin{array}{c}\overset{\displaystyle H}{\underset{\displaystyle H}{\overset{|}{\underset{|}{C}}}}-\overset{\displaystyle H}{\underset{\displaystyle Cl}{\overset{|}{\underset{|}{C}}}}\end{array}\right]_n$$

Poly(vinyl chloride)

It is a hard resin so a plasticiser was required to reduce its T_g. As with the soft acrylic materials, the plasticiser leached out in the mouth resulting in hardening of the lining. Attempts were made to reduce the hardening effect by using the copolymers of vinyl chloride and vinyl acetate (to reduce T_g) with various plasticisers [3.2.26, 3.2.27]. None were particularly successful and there were also problems with adhesion to the denture base.

$$
\left[\begin{array}{c} \text{H} \\ | \\ \text{C} - \text{C} - \\ | \\ \text{C}=\text{O} \\ | \\ \text{CH}_3 \end{array} \right]_n
$$

Poly(vinyl acetate)

Such materials are not now used as soft lining materials. A copolymer of ethylene and vinyl acetate is elastomeric and was suggested for use as a soft lining material [3.2.28] but it cannot be formed and cured using conventional dental methods.

3.2.1.7
Hydrophilic Acrylic Polymers

When hydroxyethyl methacrylate is polymerised with a small amount of cross-linking agent (e.g. ethylene glycol dimethacrylate) in the absence of solvents a hard brittle polymer results. When this is then immersed in water, it gradually swells to become soft with a final water content of ~ 37 % [3.2.29]. Such materials have been used as soft lining materials [3.2.30] but had a number of problems. These included control of final dimensions and dimensional stability due to rapid loss/gain of water.

$$
\left[\begin{array}{c} \text{CH}_3 \\ | \\ \text{CH}_2=\text{C} - \\ | \\ \text{COOC}_2\text{H}_4\text{OH} \end{array} \right]_n
$$

Poly(hydroxyethyl methacrylate)

3.2.1.8
Polyurethanes

Such materials have been used successfully in the fabrication of facial prostheses and a preliminary study indicated that they may be useful as a soft lining materials. However, there is concern over the toxicity of the isocyanates used in the production of these materials. They are discussed more fully in Sect. 3.2.3.2.

3.2.2
Tissue Conditioners

This type of material is meant for temporary usage and, as mentioned previously, should not be confused with room temperature polymerising soft acrylic materials also termed "temporary". Their use enables dentures to be worn while the oral mucosa recovers from inflammation due to ill fitting dentures or after surgery; they can also be used as temporary lining materials. A secondary use is as functional impression materials [3.2.31]. They are also used to deliver anti-fungal agents in the treatment of denture induced stomatitis [3.2.32 – 3.2.34].

They are generally presented as a powder and liquid system composition as shown in Table 3.2.5. Note there is no monomer in the liquid. When the powder and liquid are mixed, the ethanol swells the polymer beads and allows penetration by the ester plasticiser. As a result a gel is formed by polymer chain entanglement [3.2.35], the gel being essentially a solution of the polymer in the plasticiser. The resulting gel is viscoelastic in that it responds elastically to the rapid dynamic loading associated with mastication but will flow under constant fluid loads [3.2.36]. In fact it acts like a Maxwell fluid.

In the mouth, first the ethanol and then the plasticiser is lost, resulting in hardening of the material [3.2.37, 3.2.38]. It is necessary that, when used as a tissue conditioner, it should be replaced every 2 – 3 days. However, as a temporary reline material it can be used for up to 3 months, depending on the material. As a functional impression material it is recommended that it be in place for at least 24 h [3.2.39]. Bonding to the denture is good which makes them difficult to remove, a problem if regular replacement is required. Some materials do contain an anti-tack agent (e. g. Coe Comfort) making them easier to remove.

The leached ethanol can cause further irritation of traumatised tissues, especially in the high ethanol content materials. Leached plasticiser is a cause for concern over possible toxicity as with the soft acrylic, especially as the plasticiser content is higher. The leached plasticiser can also cause problems by plasticising adjacent denture base materials. This will reduce rigidity and make the denture more prone to fracture. The problem is worse when the adjacent material is of the room temperature curing type commonly used to repair dentures.

Gelation rate of tissue conditioners has been found to be influenced by molecular weight and particle size of polymer powder, amount of ethanol,

Table 3.2.5. Composition of a tissue conditioner material

Powder	Liquid
Poly(ethyl methacrylate) or related copolymer e. g. with methyl or butyl methacrylate	Aromatic ester, usually a phthalate e. g. butyl phthalyl butyl glycollate or dibutyl phthalate Others used include benzyl benzoate and benzyl salycilate + 6 – 30 % ethanol

molar volume and type of plasticiser and powder/liquid ratio [3.2.40]. It is thus possible to produce materials covering a wide range of rates and compliance.

3.2.3
Maxillo-Facial Prosthesis Materials

This group of materials is used to correct facial defects caused by cancer surgery or accidents and to repair congenital defects. Most of the materials already described for use as soft lining materials can also be used for this purpose.

Both rigid and soft acrylic materials can be used although these generally suffer from poor aesthetics. More recently, the use of terpolymer latexes [3.2.41] has been investigated. They comprise a mix of two acrylate latexes with formaldehyde as a cross-linking agent. The resulting materials have suitable properties for use as extra-oral prostheses, the prostheses are produced by dip casting over a male model.

Various silicone rubber materials are also used. RTV types [3.2.42] are based on an addition reaction using chloroplatinic acid as a catalyst, similar to the addition silicone impression materials. The heat cured materials [3.2.43] use a peroxide initiator (usually 2,4-dichloro benzoyl peroxide which requires a heat cure of 220 °C) with a vinyl terminated poly dimethyl siloxane; cross-linking occurs across the vinyl groups. The heat cured is stronger than the RTV type. The RTV silicone materials are preferred in that they most closely match the stress/strain characteristics of natural tissue.

Plasticised poly(vinyl chloride) (PVC) is also widely used [3.2.44]. It is supplied as finely divided PVC in a solvent. This is then heated and at a critical temperature the PVC will dissolve. On cooling an elastic solid is produced from which the prosthesis is processed using metal moulds at a temperature of 150 °C. They have a good combination of properties but are costly to fabricate and they lack durability.

Other materials are also used as discussed below.

3.2.3.1
Silphenylene Elastomers

Silphenylene elastomers [3.2.41] comprise a three-component RTV system. A base of poly(tetramethylsulphenylenesiloxandimethyl siloxane) with a cross linking agent of tetra propoxy silane and an organo-tin catalyst. Cross-linking occurs across the hydroxyl groups on the polymer chain ends:

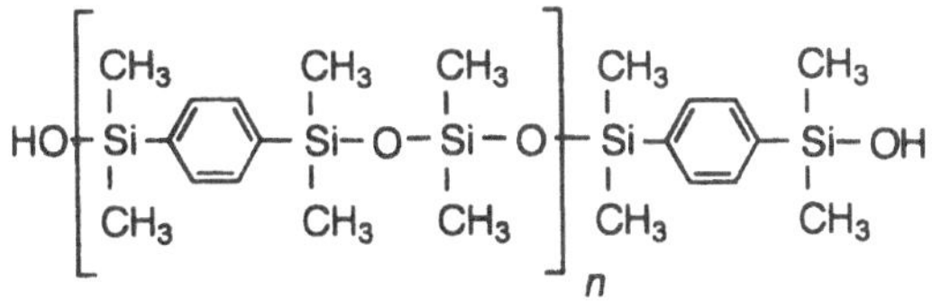

These materials have a high tensile strength but low modulus. They are transparent even when filled with silica, so can be pigmented easily for an aesthetic appearance. However, they still suffer from poor tear resistance, although this can be improved with the incorporation of modified fillers.

3.2.3.2
Polyurethanes

It was mentioned previously that polyurethanes [3.2.45] had been used success-fully for maxillo-facial prostheses. They are formed by the addition of a diiso-cyanate to a polyol in the presence of an organo-tin initiator in a dry atmosphere or carbon dioxide.

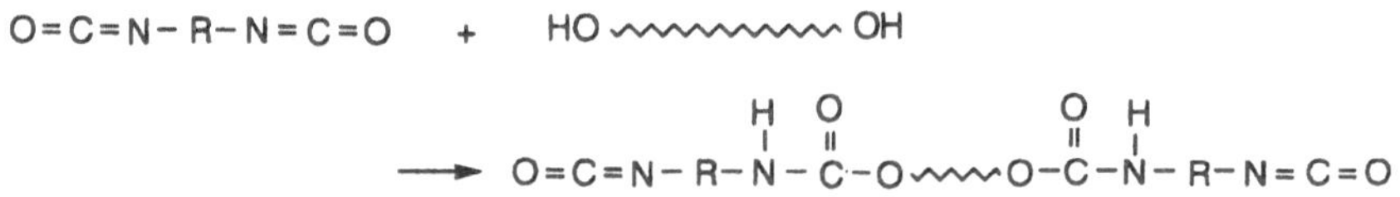

$$O=C=N-R-N=C=O \quad + \quad HO \sim\sim\sim\sim\sim OH$$

$$\longrightarrow O=C=N-R-\overset{H}{\underset{|}{N}}-\overset{O}{\underset{||}{C}}-O\sim\sim\sim O-\overset{O}{\underset{||}{C}}-\overset{H}{\underset{|}{N}}-R-N=C=O$$

R = alkyl group

Processing is carried out at 100 °C in stone moulds. Toxicity of the diisocyanate obviously limits their use to extra-oral prostheses, even then irritation can occur. Good aesthetics can be produced, but quality can be variable as the production conditions need careful control.

An isophorone polyurethane [3.2.41] is also used as a maxillo-facial material. It comprises isophorone diisocyanate, a butane diol and a polyether polyol which undergo a controlled combination to produce a prepolymer. The pre-polymer is then combined with a triol as a cross-linking agent and an organo-tin activator and processed as the other polyurethanes. The material produced is stronger than the conventional polyurethanes.

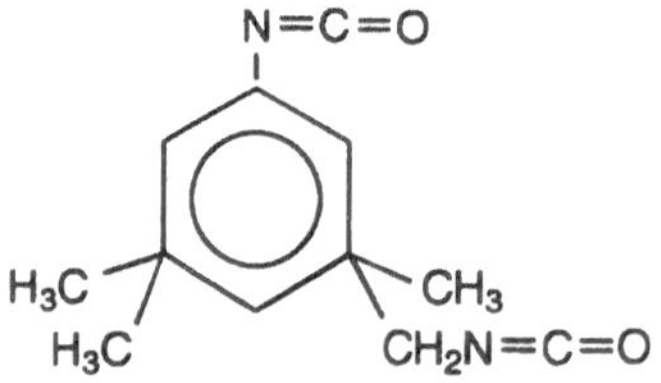

Isophorone diisocyanate

3.2.3.3
Chlorinated Polyethylene

This material [3.2.46] is prepared from industrial chlorinated polyethylene.

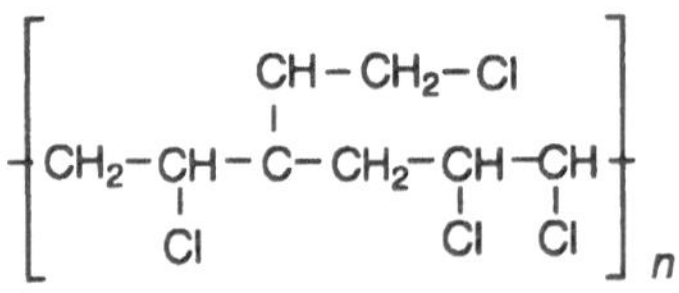

Pigments can be incorporated by compounding on a heated rubber mill to pro-duce thin sheets of material. Prostheses are fabricated using metal moulds at temperatures of up to 190 °C. More recently it has been possible to process at a lower temperature using gypsum moulds in a steam autoclave.

Chlorinated polyethylene materials have good mechanical properties and, more notably, good tear resistance.

3.2.4
Physical Properties

3.2.4.1
Water Uptake Characteristics

This is obviously an important property to evaluate where materials have to function in aqueous environments. Excessive water uptake can cause distortion and allow ingress of micro-organisms. Another potential problem is leaching of soluble matter with possible toxic effects. Ideally those materials to be bonded to PMMA should have a similar water uptake (~ 2%) [3.2.47].

Monitoring the water uptake of plasticised soft acrylic materials is complicated by gradual leaching of the plasticiser. A false equilibrium is reached where the loss of plasticiser and uptake of water is balanced. Figure 3.2.6 shows the example of a plot of percent water uptake of a plasticised soft acrylic material and a representative sample of other commercial soft lining materials [3.2.5].

Water uptake of elastomers can be high and prolonged, due to the presence of water soluble or hydrophilic components. The experimental polymerisable plasticiser material reported earlier proved to have high and prolonged water uptake (see Fig. 3.2.7). During clinical trials it was thought that this caused bubbling to form on the surface of the lining (see Fig. 3.2.5). Water soluble impurities were proposed as the possible cause of this high water uptake [3.2.7]. As the water diffuses into the material it will reach impurity sites and a solution droplet is formed. The droplet will then grow until the osmotic pressure is balanced by the elastic forces of the material. The driving force is the chemical potential gradient between the external solution and the internal solution

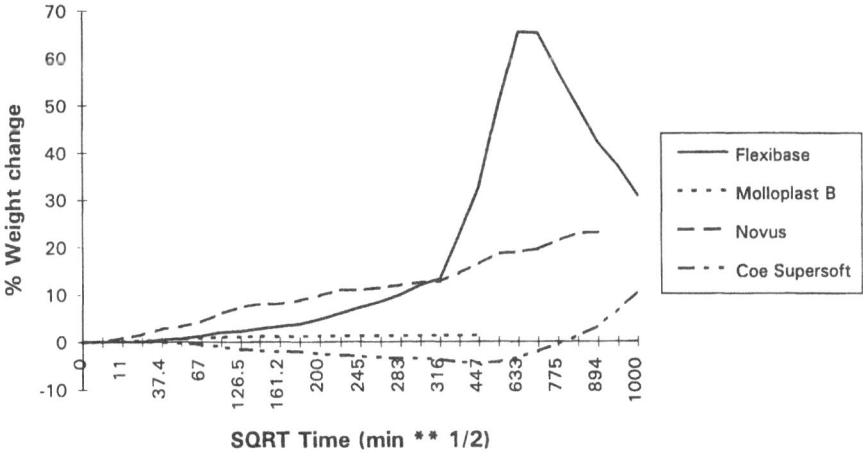

Fig. 3.2.6. Water absorption characteristics of some soft lining materials

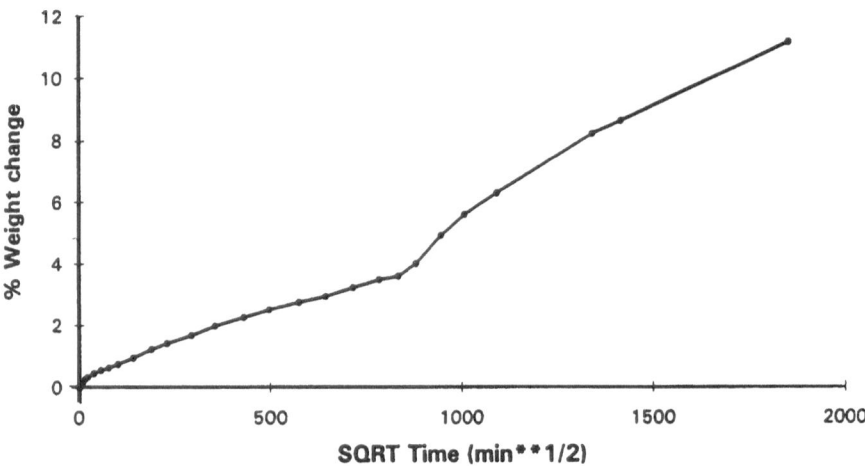

Fig. 3.2.7. Water uptake of an experimental soft lining material

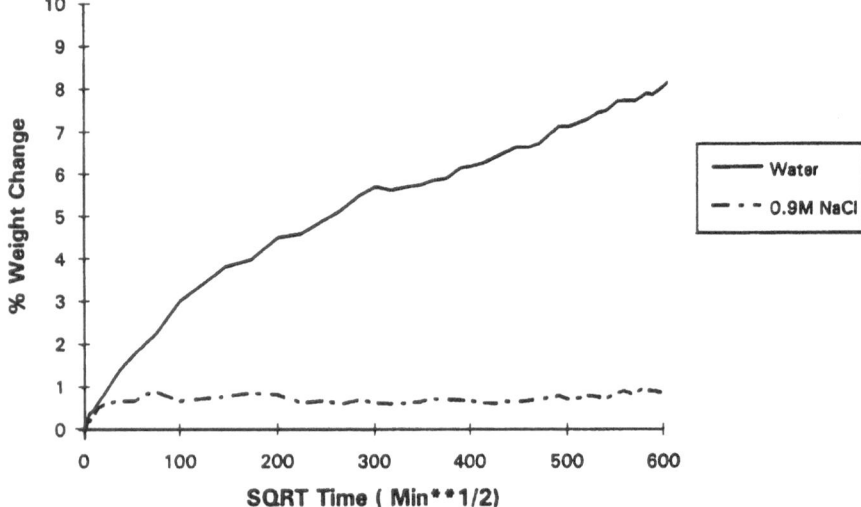

Fig. 3.2.8. Water uptake of an experimental soft lining material from distilled water and 0.9 mol/l saline

droplet. If the external solution is distilled water the gradient will be large, giving rise to high water uptake. However, if saline (or other ionic) solution is used the gradient will be lower, resulting in lower water uptake. Figure 3.2.8 shows water uptake of an experimental elastomer/methacrylate monomer system in distilled water and in 0.9 mol/l saline [3.2.48]. Note the reduced uptake of the material in the saline solution.

In summary, long term water uptake of elastomers consists of an advancing front of growing water droplets (see Fig. 3.2.9). The presence of these strained

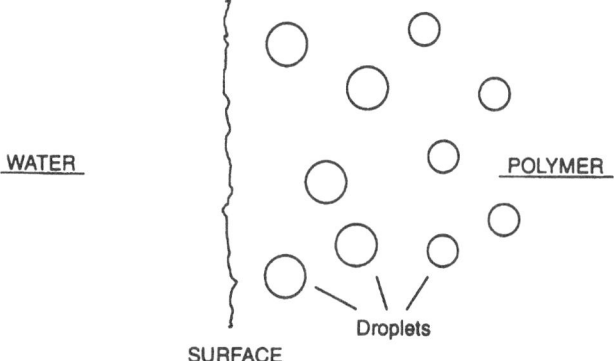

Fig. 3.2.9. Schematic representation of the water uptake mechanism of elastomeric materials

sites will drastically reduce the strength of the material. In weak materials (such as the experimental polymerisable plasticiser material mentioned previously) osmotic pressures within the droplets can be high enough to cause rupture.

Water uptake will depend upon several aspects:

1. intrinsic water uptake of the material;
2. mechanical properties;
3. osmotic pressure of external solution; and
4. presence of soluble/hydrophilic components within the material.

It is thus important to measure mechanical properties in a saturated state as this is the state in which the materials are expected to function.

3.2.4.2
Viscoelastic Properties

All the materials discussed in this chapter are viscoelastic in that their properties are time dependent. Relevant properties are shear modulus (G) and loss tangent ($\tan\delta$). Table 3.2.6 gives values for some soft lining materials. Shear modulus values give an indication of compliance and $\tan\delta$ of resilience as shown in Eqs. (3.2.1) and (3.2.2)

$$\text{Compliance} = 1/\text{modulus} \tag{3.2.1}$$

$$\text{Resilience} = 1 - e^{-\pi\tan\delta} \tag{3.2.2}$$

Resilience values are often quoted as a percentage and the $\tan\delta$ values shown in Table 3.2.6 give rise to resilience values of 4–85%, silicone rubber based materials having the highest values and plasticised acrylic materials the lowest. With the currently available materials covering such a large range it is not possible to determine what value is required clinically. A perfectly elastic material would have a $\tan\delta = 0$ which is equivalent to a resilience value of 100%.

Table 3.2.6. Viscoelastic properties of some soft lining materials at 37 °C [3.2.49]

Material	Type	Shear Modulus (GPa)	$\tan\delta$
Coe Supersoft	Plasticised, heat cured acrylic	6.8	0.9
Flexibase	Condensation silicone	8.8	0.08
Molloplast B	Modified, heat cured silicone	5.5	0.05

Table 3.2.7. Peel energies of some soft lining materials [3.2.53]

Material	Peel energy ($KJ\ m^{-2}$)		
	Dry	7 days in water	90 days in water
Coe Supersoft	13.39	13.69	14.97
Molloplast B	1.80	2.23	1.98
Flexibase	0.63	1.19	0.3

3.2.4.3
Adhesion to PMMA

Generally, silicone materials have poor adhesion and soft acrylic materials have good adhesion [3.2.8]. Newer materials of both have improved bonding, sometimes by the use of an adhesive or adhesion promoter [3.2.50].

Testing of adhesion is best done using a peel test first described by Kendall [3.2.51] and used by Wright [3.2.8], Parker [3.2.52] and Sinobad et al. [3.2.53]. It

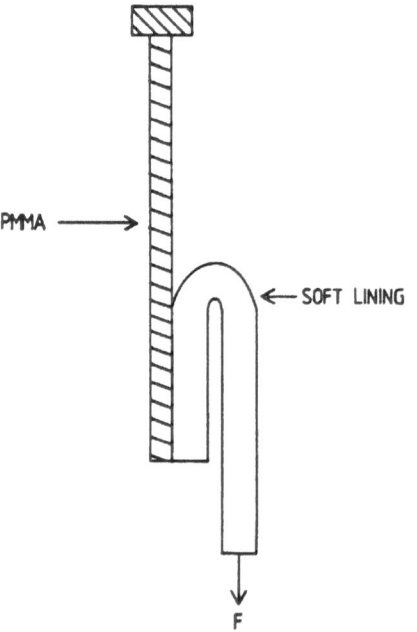

Fig. 3.2.10. Peel test specimen as tested

PMMA →

← SOFT LINING

F

is sometimes difficult to ensure that failure of the bond occurs at the interface, although failure within the lining material is common. Adhesion can also be assessed by a tensile test, although failure of the liner is still a common problem. Tests should also be carried out after immersion in water as this can have a damaging effect on the bonding of some of the materials. Table. 3.2.7 gives some representative values obtained using a peel test with a peel angle of 180° as shown in Fig. 3.2.10. It should be noted that values will depend on testing rate.

3.2.4.4
Stress/Strain Characteristics

All such materials are non-linear in tension except near the origin (see Fig. 3.2.11). Table 3.2.8 gives typical values for ultimate tensile strength and elongation to break. Energy to break is the area under the stress/strain curve and so is influenced by both these parameters. Materials represented in Fig. 3.2.11 have similar tensile strengths but the experimental elastomer material has the highest energy to break value. Again properties are rate dependent.

3.2.4.5
Rupture Properties

An important property for soft lining materials as this type of force is the one they will most commonly experience in use. They are best assessed as tear energies using a "trouser" test piece (see Fig. 3.2.12) first used for rubbers by Rivlin and Thomas [3.2.55] and adapted for soft lining materials by Parker and Braden [3.2.56], Wright [3.2.58] and Dootz et al. [3.2.54]. Table 3.2.9 gives representative values. Generally, silicone rubber based materials have lowest tear energy; this can lead to tearing of the lining at the periphery of the denture.

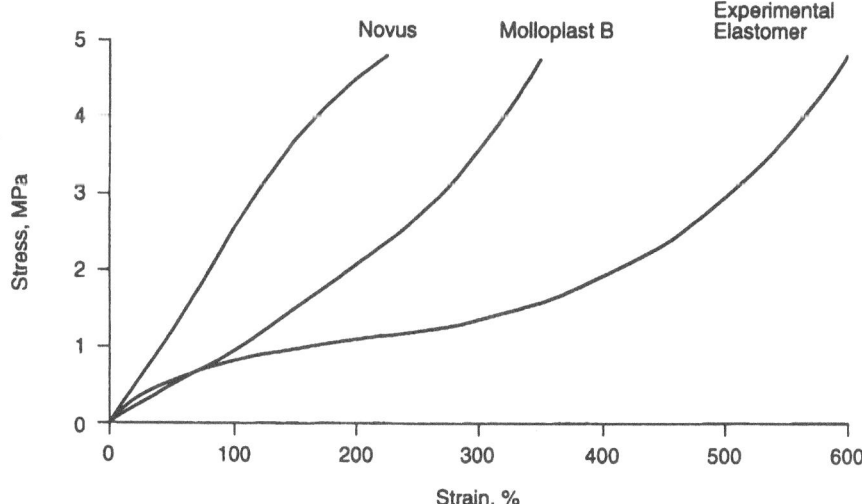

Fig. 3.2.11. Stress/strain curves of some soft lining materials

Table 3.2.8. Tensile properties of some soft lining materials [3.2.54]

Material	Tensile strength (MPa)	Elongation at break (%)
Coe Supersoft	2.66	230
Molloplast B	4.23	325
Novus (A polyphosphazine)	3.60	240

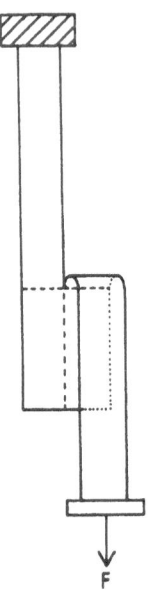

Fig. 3.2.12. Trouser tear test specimen as tested

Table 3.2.9. Tear energies of some soft lining material [3.2.54]

Material	Tear energy (KJ m^{-2})
Coe Supersoft	11.5
Molloplast	1.43
Novus	21

3.2.4.6
Wettability

A requirement of soft lining materials is that they are wetted by saliva, as a thin layer of saliva is required to aid retention of the denture and reduce frictional forces on the oral mucosa [3.2.58]. Wettability is assessed by measurement of contact angles (see Fig. 3.2.13) where angles < 90° are non-wetting, and angles > 90° are wetting.

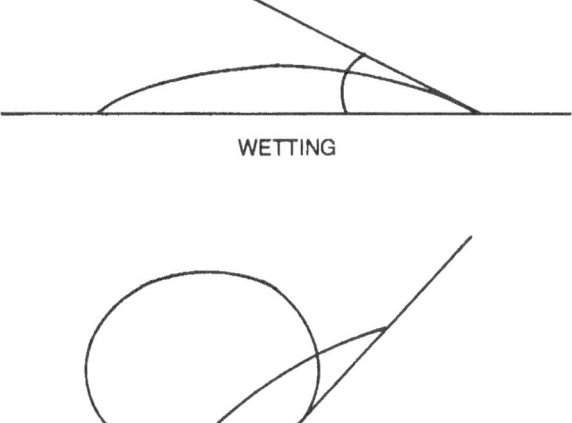

WETTING

NON-WETTING

Fig. 3.2.13. Wetting and non-wetting contact angles

Table 3.2.10. Contact angles of some soft lining materials [3.2.58]

Material	Contact angle (degrees)	
	As processed	After 6 months in water at 37 °C
Coe Supersoft	69	49.5
Flexibase	90.5	71
Molloplast B	82.5	81

Some silicone rubber and natural rubber materials have poor wettability. In some materials wettability is improved after water immersion [3.2.13]. Table 3.2.10 gives some representative values.

3.2.5
References

3.2.1. Sartomer Monomer (1971). Technical Bulletin SR 233
3.2.2. Lee WA, Rutherford RA (1975). In: Brandup J, Immergut EH (eds) Polymer Handbook, 2nd edn. John Wiley, New York, III-139
3.2.3. Aiken A (1988) Ph.D. Thesis. University of London, London
3.2.4. Autian J (1973) Envir. Health Perspectives Jun: 3
3.2.5. Braden M, Wright PS (1983) J. Dent Res 70:210
3.2.6. Litchfield J, Wood LG (1965) British Patent 983 817
3.2.7. Parker S, Braden M (1989) Biomaterials 10:91
3.2.8. Wright PS (1982) J Dent Res 61:1002
3.2.9. Wright PS (1980) J Dent 8:144
3.2.10. Rochow ER (1987) Silicon and Silicones. Springer, Berlin Heidelberg New York p 109

3.2.11. Van Handel AB (1974) US Patent 3 785 054
3.2.12. Beers MD (1968) US Patent 3 382 205
3.2.13. Wright PS (1980) PhD Thesis. University of London, London
3.2.14. Segal BW, Glassman A. (1982) J Prosthet Dent 47:85
3.2.15. Parker S (1994) Abstract 1384 J Dent Res 73 (special issue)
3.2.16. Nishiyama M, Kato T (1987) J Nihon Univ Sch Dent 29:100
3.2.17. May P, Guerra LR (1981) US Patent 4 251 215
3.2.18. Gettleman L, Rappoport L, Marx SA, Watters, JC, Selvaraj R, Hazelrigg D, Khan Z (1994) Abstract 1382 J Dent Res. 73 (special issue)
3.2.19. Parker S, Braden M (1990) Biomaterials 11:482
3.2.20. Parker S (1993) J Dent Res 72 (special issue)
3.2.21. Parker, S, Riggs, P, Martin, D and Kalchandra, S (1995) Abstract 1676 J Dent Res 74 (special issue)
3.2.22. Hayakawa I, Kawae M, Tsuji Y, Masuhara E (1984) J Prosthet Dent 51:310
3.2.23. Ohe Y, Kadoma Y, Imai Y (1990) Shika-Zairyo-Kikai 9:654
3.2.24. IARC (1976)Monographs on the evaluation of the carcinogenic risks of chemicals to man 12:262
3.2.25. Matthews E (1945) Brit Dent J 78:140
3.2.26. Bains MED (1957) MSc Thesis. University of Manchester, Manchester
3.2.27. Bates JF (1965) J Amer Dent Assoc 70:344
3.2.28. Nishiyama M (1972) J Jap Res Soc Dent Mat Appliances 27:120
3.2.29. Simpson BJ (1969) Biomed Med Eng 4:65
3.2.30. Sklover IT, Tendler MD (1967) Dent Dig 73:45
3.2.31. Wilson HJ, Tomlin HR, Osborne J (1966) Brit Dent J 121:9
3.2.32. Douglas WH, Walker DM (1973) Brit Dent J 135:55
3.2.33. Thomas CJ, Nutt GM (1978) J Oral Rehab 5:167
3.2.34. Shneid TR, Rawls, HR, Bradley DV (1992) Abstract 1561 J Dent Res 71 (special issue)
3.2.35. Braden M (1970) J Dent Res 49:496
3.2.36. Braden M (1971) Rheol Acta 10:86
3.2.37. Braden M, Causton BE (1971) J Dent Res 50:1544
3.2.38. Jones DW, Sutow EJ, Hall GC, Tobin WM, Graham BS (1988) Dent Mater 62:421
3.2.39. Graham BS, Jones DW, Sutow EJ (1989) J Prosthet Dent 62:421
3.2.40. Murata H, Shigeto N, Hamada T (1992) Abstract 427 J Dent Res 71 (special issue)
3.2.41. Lewis DH, Castleberry DJ (1980) J Prosthet Dent 43:426
3.2.42. Moore DJ, Glaser ZR, Tabacco MJ, Linebaugh MG (1977) J Prosthet Dent 38:319
3.2.43. Lontz JF (1990) Dent Clin North Amer 34:307
3.2.44. Sweeney WT, Fischer TE, Castleberry DJ, Cowperthwaite BS (1972) J Prosthet Dent 27:297
3.2.45. Gonzalez JB, Chao EYS, An K-N (1978) J Prosthet Dent 39:307
3.2.46. May PD (1978) J Biomed Mat Res 12:421
3.2.47. Stafford DG, Braden M (1968) J Dent Res 47:341
3.2.48. Parker S, Riggs P, Kalachandra S, Braden M, Taylor D (1995) Abstract 1474 J Dent Res 74 (special issue)
3.2.49. Braden M, Clarke RL (1972) J Dent Res 51:1525
3.2.50. McMordie R, King GE (1989) J Prosthet Dent 61:636
3.2.51. Kendall K (1971) J Phys D Appl Phys 4:1186
3.2.52. Parker S (1982) PhD Thesis. University of London, London
3.2.53. Sinobad D, Murphy WM, Hugget R, Brooks S (1992) J Oral Rehab 19:151
3.2.54. Dootz ER, Koran A, Craig RG (1992) J Prosthet Dent 67:707
3.2.55. Rivlin RS, Thomas AG (1953) J Polym Sci 10:291
3.2.56. Parker S, Braden M (1982) J Dent 10:149
3.2.57. Wright PS (1980) J Dent Res 59:614
3.2.58. Wright PS (1981) Proc Europ Prosthodont Assoc 4:134

Sensitive systems

V. Shibaev (Ed.)

Polymers as Electrooptical and Photooptical Active Media

1996. XVII, 211 pages. 117 figures, 30 tables.
Hardcover DM 148,–
ISBN 3-540-59486-8

Polymeric materials have special advantages over other materials used for the recording, storage and retrieval of information, telecommunication transmission and visualization of images. The authors describe the synthesis, the physicochemical behavior and the applications of these highly sensitive macromolecular systems. They discuss the most essential developments in this field.
For scientists and professionals working in the field of electrooptical and photooptical polymeric materials.

Prices subject to change without notice.
In EU countries the local VAT is effective.

Please order by
Fax: +49 - 30 - 827 87 - 301
e-mail: orders@springer.de
or through your bookseller

Springer

Springer-Verlag, P. O. Box 31 13 40, D-10643 Berlin, Germany

New journal

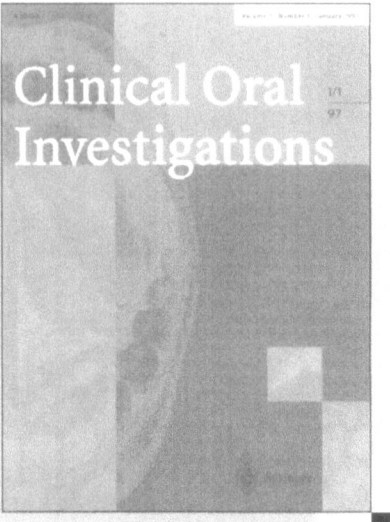

Clinical Oral Investigations
founded by the
German Society for
Oral and Maxillofacial
Medicine (DGZMK)

Editor-in Chief:
G. Schmalz

ISSN 1432-6981
Title No. 784

Clinical Oral Investigations
publishes original articles and
invited reviews. The journal aims
to provide an international
readership with up-to-date results
of basic and clinical studies in the
studies in the field of oral and
maxillofacial science and medicine and to clarify the relevance of
these findings for a modern
practice. Topics covered are:
Maxillofacial and Oral Surgery,
Prosthetic and Restorative
Dentistry, Endodontics,
Perodontology, Orthodontics,
Dental Material science, Clinical
Trials, Epidemiology, Preventive
Dentistry, Oral Pathology, Oral
Implantology, and Oral Basic
Sciences.

Subscription information
for 1997:
Volume 1 (4 issues)
DM 296,–*

Please order by
Fax: +49 - 30 - 827 87 - 448
e-mail: subscriptions@springer.de
or through your bookseller

* plus carriage charges.
In EU countries the local VAT is effective.

Springer

Springer
and the
environment

At Springer we firmly believe that an international science publisher has a special obligation to the environment, and our corporate policies consistently reflect this conviction.

We also expect our business partners – paper mills, printers, packaging manufacturers, etc. – to commit themselves to using materials and production processes that do not harm the environment. The paper in this book is made from low- or no-chlorine pulp and is acid free, in conformance with international standards for paper permanency.

 Springer